TIME SERIES MODELLING IN EARTH SCIENCES

Time Series Modelling in Earth Sciences

B.K. Sahu
Department of Earth Sciences
IIT Bombay, Mumbai
India

A.A. BALKEMA PUBLISHERS LISSE/ABINGDON/EXTON (PA)/TOKYO

Published by: A.A. Balkema, a member of Swets and Zeitlinger Publishers
www. balkema.nl and www.zsp.swets.nl

ISBN 90 5809 267 4

Printed in India

To
My wife Kanchan ...

PREFACE

This book represents materials culled from my Mathematical Geology and Geomodelling courses, respectively for Ph.D. and M.Tech. students of Geology at IIT Mumbai for the past 23 years. Most of the geological data are stochastic in nature and since our Earth is a dynamic system evolving in time, 'Time Series Modelling' seems most appropriate for a logical interpretation of the spatial and spatio-temporal data which are essentially dependent on previous values (samples). However, as most of the geology students lack a mathematical and statistical background, they should have a basic knowledge for a statistical analysis of independent data (sample) using univariate/multivariate models (see Sahu, B.K., 2002, Statistical Models in Earth Sciences, in press) that has been covered in M.Sc. (Applied Geology) programme at IIT Mumbai.

This text introduces simpler Linear Time Series models which can be seasonal or non-seasonal (stationary or non-stationary), modelled as SARIMA or ARIMA. Later, non-linear time series models such as sudden-bursts, threshold/exponential autoregressive, state-space models are used to solve more complex and realistic Earth Sciences problems. Several real geological examples/case studies have been included which exemplify realistic modelling problems in Earth Sciences. In addition, several exercises and examples from other fields are also included which are useful for modelling earth sciences time (spatial) series observations. It is hoped that this book will be useful to students, faculty and professionals in Earth Sciences and as well as for other related scientific/engineering fields.

I thank CSIR, New Delhi for their financial assistance through the award of Emeritus Professorship for the tenure of writing this entire book. I thank the authorities of IIT Mumbai, especially the Department of Earth Sciences for providing me with a congenial environment and suitable infrastructural assistance in order to complete this book in time. I also thank my former M.Sc., M.Tech., and Ph.D. students who have contributed to my pursuit of realistic time series modelling in Earth Sciences. I specially thank Shri Bhaskar Chandra Sahoo, JRF, for diligently converting the handwritten manuscript to an excellent computer softcopy. I shall greatly appreciate if any mistakes and errors are brought to my notice for future corrections.

March 2002   **B.K. Sahu**

CONTENTS

INTRODUCTION

1.1 DYNAMIC EARTH

The Earth is a dynamic planet with optimal size as well as distance from Sun, the centre of its revolution. The optimal distance from the Sun enables Earth to obtain adequate heat for the development and sustenance of organic life at its surface, hydrosphere and regolith. Simultaneously, it is not near enough to the Sun to lose its atmosphere and hydrosphere, which generate the deterministic climatic seasons, depending on its position in its revolution path.

The Earth is large enough to have its gravity field together with internal energies of heat resulting from chemical reactions, mechanical processes and radioactive decay of radioactive elements. Internal processes are manifested through irregular tectonic cycles along with associated basinal, magmatic and metamorphic cycles.

We are well acquainted with different surficial processes such as sediment transport/deposition by rivers, wind action, oceanic waves and currents, and glaciers. These surficial processes are due ultimately to gravity and/or the Sun's heat energy. However, dynamic movements due to the Earth's internal energy are not visulized at the surface but its effects are seen later, in terms of volcanic eruptions, earthquakes, tectonic movements resulting in basin and mountain fromation, folding/faulting, etc. Therefore, the surficial and internal dynamic processes together form the components of the Earth systems, which contain many sub-systems of interest in Earth sciences (Tables 1.1, 1.2 and Fig. 1.1).

Table 1.1 Common forms of energy and then associated activities

Energy	Activity in the Earth
Kinetic	Waterflow, wind, waves, landslides
Heat	Rain/snowstorms, volcanoes, hotsprings
Chemical	Decay of vegetation, forest fires, burning coal/oil, phase change
Electrical	Lightning, aurora
Radiation	Daylight, hydrologic cycle
Atomic	Heat in the Earth's crust and interior
Impact	Heat from meteorite impact

Table 1.2 Internal and external structure (layers) of the Earth

			State	Depth/Height (in km)	Nature
I		Lithosphere	Solid	0-100	Cool and rigid
N	M	Asthenosphere	Plastic	100-350	Hot, weak, plastic
T E R N A L	A N T L E	Mesosphere	Solid	350-2883	Hot but strong (high pressure)
		Outer Core	Liquid	2883-5140	Liquid iron
		Inner Core	Solid	5140-6371	Solid iron
E X		Hydrosphere	Liquid	0-10	Water, snow, ice
T E		Atmosphere	Gas	0-200	Gases of H_2O, O, N, etc.
R N		Biosphere	Organic	-	Life forms
A L		Regolith	Solid	0-2	Soils, laterites

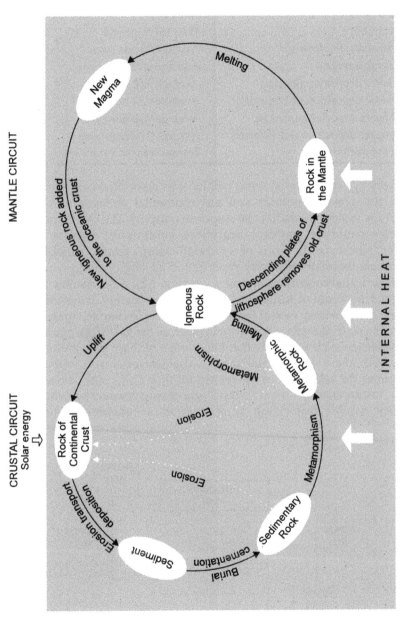

 Fig.1.1 External and internal processes affecting the various stages of rock cycle in the dynamic earth system

Some dynamic characteristics of the Earth

Geothermal gradient	15 to 75°C/km
Heat flow	Average 6.3×10^{-6} J/m^2/sec
Plate motion	15 to 161 mm/year
Sea level rise (Holocene)	Average 20 mm/year
Basin subsidence	Average 0.10 mm/year
Siliclastic deposition (glacial)	Average 0.05 mm/year
Siliclastic deposition (interglacial)	Average 0.12 mm/year
Carbonate deposition (glacial)	Average 0.03 mm/year
Carbonate deposition (interglacial)	Average 0.09 mm/year
Peat deposition	2.5 mm/year to 5 mm/year

Every system (or subsystem) will have its associated environment in which the system operates and environmental changes may cause (input) the system to modify the response (output). The cause and effect variables could be physical/analytical, continuous/discrete, deterministic/stochastic. In terms of interrelationships among variables, the systems can be mathematically represented as single/multiple input(s) with their initial conditions at time $t = t_0$ (Fig. 1.2). Also systems may be described in terms of lumped/distributed parameters, constant/time-varying parameters, causal/non-causal, static/dynamic, linear/non-linear. The state of a dynamical system (such as the Earth) is a minimal set of variables in such a manner that knowledge of these variables at $t = t_0$, along with the values of the input variables for $t > t_0$, completely determines the behaviour of the system for time $t > t_0$. Static systems are called instantaneous, memoryless, whereas dynamic systems with memory exhibit interesting behaviour as a function of time. In

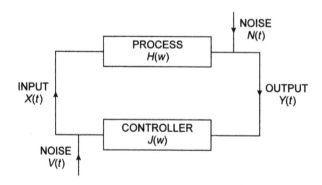

Fig. 1.2 Closed-loop system

vector-matrix notation, we can write

$$\dot{x}(t) = f(x(t), r(t)) \qquad \text{state equation ...(1)}$$

where x is a n-dimensional state row, r is a m-dimensional input vector and f is a n-dimensional function vector. The output $y(t)$ is a function of initial state $x(t)$ and input $r(t)$ given by

$$y(t) = g(x(t), r(t)) \qquad \text{output equation ...(2)}$$

These two equations form the state model of the system.

However, the output equations may have explicit dependence in time t as well (time-varying systems) and these equations will then become

$$\dot{x}(t) = f(x(t), r(t), t) \qquad \text{state equation ....(3)}$$

$$y(t) = g(x(t), r(t), t) \qquad \text{output equation ...(4)}$$

Note that output equation is not dynamic but a static (instantaneous) one. To find the solution to the state model, we need to solve the system state equation. If inputs $r(t)$ be constant, then

$$\dot{x}(t) = f(x(t))$$

Consider state equation $\qquad \dot{x}(t) = f(x(t), r(t), t) \qquad$ ...(3)

If $x(t)$ is the solution with initial condition $x(t_0)$ at $t = t_0$ and input $r(t)$, $t > t_0$,

i.e. $\qquad x(t) = \phi(x(t_0), r(t))$

Linearity implies homogeneity $\phi(x(t_0), \alpha r(t)) = \alpha \phi(x(t_0), r(t))$ and superposition, i.e. $\phi(x(t_0), r_1(t) + r_2(t)) = \phi(x(t_0), r_1(t)) + \phi(x(t_0), r_2(t))$

Thus, linearity implies $\phi(x(t_0), \alpha r_1(t) + \beta r_2(t)) = \alpha \phi(x(t_0), r_1(t)) + \beta \phi(x(t_0), r_2(t))$

i.e. linearity = homogeneity + superposition

A system is *linear* in nature if it satisfies decomposition property and the decomposed components are linear. We define the term decomposition below.

Let $x'(t)$ be the solution to Eq. (3) when the system is in zero state for all inputs $r(t)$,

i.e. $x'(t) = \phi(0, r(t))$ and $x''(t)$ be the solution when for all states $x(t_0)$ the input $r(t)$ is zero,

i.e. $x''(t) = \phi(x(t_0), 0)$

The system has decomposition property if

$$x(t) = x'(t) + x''(t)$$

or $\qquad \phi(x(t_0), r(t)) = \phi(0, r(t)) + \phi(x(t_0), 0)$ ...(5)

Thus, a system is linear provided it satisfies the decomposition property and has zero input linearity and zero state linearity. Otherwise, the system is non-linear. Zero input linearity refers to when input $r(t) = 0$, the zero input response satisfies homogeneity and superposition with respect to initial states. Zero state linearity refers to the state when $x(t_0) = 0$ the zero state response satisfies homogeneity and superposition with respect to the inputs.

Example 1
   (i)   Linear $y(t) = ax(0) + br(t)$
   (ii) Non-linear $y(t) = (r(t))^2$ as it is not homogeneous.
   (iii) Non-linear $y(t) = ax(0) + (r(t))^2$. It has decomposition property and is zero input linear but not zero state linear.

The state-variable model for a linear time-invariant system can be given as

$$\dot{x} = Ax + Br \qquad \text{state equation ...(6)}$$

$$y = Cx + Dr \qquad \text{output equation ...(7)}$$

where A, B, C, D matrices are functions of time. The state-variable model is most useful in large systems as all the system states become visible for all time in addition to obtaining output information. However, for small size systems, differential equation models give physical insight into system behaviour through frequency domain techniques.

For discrete-time systems, we unite x_t instead of $x(t), r_t$ in place of $r(t)$ and the corresponding state-variable model becomes

$$x(k+1) = Ax(k) + Br(k) \qquad \text{state equation ...(8)}$$

$$y(k) = Cx(k) + Dr(k) \qquad \text{output equation ...(9)}$$

where k's are discrete times.

Even if some systems are non-linear, we can use linearization methods by making the non-linearity changed to piece-wise linearity, or by expanding into Taylor's series and omitting the higher order terms above the first (linear) term. Considering the output variable y_j of a multiple input system described by the static functional relation (output equation)

$$y_j = \psi_j(r_1, r_2, ..., r_m)$$

and expanding in Taylor's series and operating points ($y_{10}, ..., y_{n0}, r_{10}, ..., r_{m0}$), we obtain the following linearized model by evaluating the partial derivatives at the operating point:

$$y_j - y_{j0} = \left(\frac{\partial \psi_j}{\partial \psi_{r_1}}\right)_0 (r - r_{10}) + \left(\frac{\partial \psi_j}{\partial \psi_{r_2}}\right)_0 (r - r_{20}) + \cdots + \left(\frac{\partial \psi_j}{\partial \psi_m}\right)_0 (r - r_{m0})$$

...(10)

In case the system is static and single input-single output, equation (10) reduces to a simpler linear equation as

$$y - y_0 = \left(\frac{\partial \psi}{\partial r}\right)_0 (r - r_0)$$

...(11)

where $y = \psi(r)$.

Linearization of dynamic nonlinear systems follows similarity

$$\dot{x}(t) = f(x(t), r(t)), \qquad \text{continuous time} ...(12a)$$

$$x(k + 1) = f(x(k), r(k)) \qquad \text{discrete time} ...(12b)$$

which, on approximate linearization yields:

$$\tilde{x}(t) = A\tilde{x}(t) + B\tilde{r}(t) \qquad \text{continuous time} ...(13a)$$

$$\tilde{x}(k + 1) = B\tilde{x}(k) + B\tilde{r}(k) \qquad \text{discrete time} ...(13b)$$

where ~ indicates derivatives for operating point.

For any linear/non-linear system, we should find its equation slope x_0 which is constant under constant input r_0.

x_0 is solution of equation

$$\dot{x}(t) = f(x(t), r_0) = 0 \qquad \text{continuous time} ...(14a)$$

$$x(k + 1) - x(k) = f(x(k), r_0) = 0 \qquad \text{discrete time} ...(14b)$$

We generally use the discrete time methodology as this is amenable to easy computer storage/retrieval as well as mathematical analysis. It may be noted that any continuous time record such as temperature or pressure of air can be discretized as finely as needed in order to obtain discrete time sequences for computer storage and analysis. We also emphasize the time-domain approach as it is easily visualized compared to frequency-domain approach, which is transformed from the observed time-domain of data collection. However, we consider time-domain and frequency-domain approaches of time series analyses as complimentary models as these highlight the different aspects of the time series rather than competing models (as has been described in most of the textbooks).

Conventional statistical methods based on assumption that adjacent observations (data) are independent and identically distributed are not applicable to the time series where adjacent data tend to be highly correlated and separated data may have non-linear (linear) trends. Time series analyses address these questions posed by time correlations and time trends.

Time domain and frequency domain applications are not necessarily mutually exclusive but rather, complimentary procedures to highlight the characteristics of the generating process for the time series.

Time Domain → linear regressions of past values, forecasting using Box-Jenkins ARIMA with provision of having multiple inputs using multivariate ARIMA or with transfer function modelling.

Recalling that additive models are also being used, the resulting state-space model can be handled through Kalman filters and smoothing that were originally developed for estimation and control in space flights.

Frequency Domain → relation to periodic (or systematic sinusoidal) variables in most data. The periodicity is due to biological processes, physical processes or environmental phenomena giving climatic variations in terms of hours, days, months, quarters, years, etc. Spectral analysis partitions the power associated with each period of interest. Although for long series, the time domain and frequency domain methods give similar results (models), for short series, time domain models are more precise and preferable.

1.2 TIME/SPATIAL SERIES

Time series or one-dimensional spatial series along a line have the structure: x_t at discrete time $t = 0, 1, 2, ... N$ such that the neighbouring values may be strongly dependent (correlated), whereas values further apart in time (or spatial location) are likely to be independent (uncorrelated). Here, the random variable x may be scalar or vector valued and a suitable transformation may be necessary to make x Gaussian (x scalar) or multivariate distribution (x vector). The most common normalizing/linearizing transformation is the logarithmic transformation. However, other more specialized transformations, such as Box and Cox's power transform may be useful, if necessary.

The main objectives of the time series analysis are to model the data, to provide a compact description, to estimate the parameters of the model, to check/validate the model and possibly, to understand the generating process for the data. Time series data may contain trend (signal), seasonalities, abrupt changes, random errors and outliers. The outliers must be identified either visually or by statistical tests and eliminated/corrected. The seasonalities have fixed frequencies which are to be eliminated (filtered) before long-term trends can be estimated/interpreted. Also, there are other objectives such as predicting one time-series based on other (input) time series, forecasting and simulation studies (see Figs. 1.3, 1.4, 1.5, 1.6).

(i) *Trend and seasonality*

Let $\qquad\qquad X_t = m_t + Y_t$ $\qquad\qquad\qquad\qquad\qquad$ ...(11)

where m_t is a deterministic polynomial trend given by $a_0 + a_1 t + a_2 t^2$ and Y_t is a random noise sequence having zero mean. The parameters a_i, $i = 0$, $1, 2$ can be estimated by least squares fit to minimize sum of square errors

$$\sum_{t=1}^{n} (x_t - m_t)^2.$$

For example, if $\qquad X_t = m_t + s_t + Y_t$ $\qquad\qquad\qquad\qquad$ ...(12)

where m_t is the trend, s_t is seasonality with period d and Y_t is random noise. Then, using classical decomposition method, we have $EY_t = 0$, s_{t+d}

$= s_t$ and $\sum_{j=1} s_j = 0$.

Further, if d is odd $(= 2q + 1)$, we can use simple moving average to estimate trend as:

$$\hat{m}_t = (2q + 1)^1 \sum_{j=-q}^{q} X_{t-j} \, ; q + 1 \le t \le n - q \qquad\qquad ...(13a)$$

and if d is even $(= 2q)$, then we use the moving average to estimate trend as:

$$\hat{m}_t = \left(\frac{1}{2} x_{t-q} + x_{t-q+1} \cdots + x_{t+q-1} + \frac{1}{2} x_{t,q} \right) / d \, ; q < t \le n - q \qquad ...(13b)$$

In such a case, it is easy to estimate seasonality components as follows:

Define $w_k = \left\{ \left(x_{k+jd} - \hat{m}_{k+jd} \right), q < k + jd \le n - q \right\}$

Then, seasonality component is estimated as

$$\hat{s}_k = w_k \frac{1}{d} \sum_{t=1}^{d} w_t , i = 1, 2, 3, ..., d \qquad ...(14)$$

and it is known that $\qquad \hat{s}_k = \hat{s}_{k-d}, k > d.$

The deseasonalized series $d_t = x_t - \hat{s}_t, t = 1, 2, ..., n$ $\qquad ...(15)$

contains trend and random noise and trend m_k has to be re-estimated from Eq. (15) data by using Eq. (13a, b). The random noise is now given by the equation

$$\hat{Y}_t = x_t - \hat{m}_t - \hat{s}_t, t = 1, 2 ..., n \qquad ...(15)$$

where $\hat{m}_t$ is re-estimated trend for deseasonalized data. An alternative method of estimating trend/seasonality is through the use of back-shift operators B, B_d as proposed by Box and Jenkins (1970, 1976) and is discussed later.

A complete probabilistic time series models for sequence of random variables $\{X_1, X_2, ...\}$ specifies all joint distributions of these random vetors. It is generally sufficient to specify the first and second order moments of the joint distribution through $E(X_t)$ and $E(X_{t+h}, X_t), t = 1, 2,...;$ $h = 0, 1, 2,....$ It is well known that Gaussian distributions are characterized by first and second order cumulants, as all higher cumulants are zero. Hence, second order characterization of time series models is generally sufficient, provided X_t is Gaussian, i.e. weak stationarity is equivalent strict stationarity for Gaussian process. We give a few examples of forward modelling of some zero mean process.

Example 1   Independent and identically distributed noise (IID)
Each x_i is independent and identically distributed noise. Thus, $E(x_i) = 0$, $E(X_i, X_j) = 0, i \neq j$ and $= \sigma^2 < \infty, i = j$, i.e. $\{X_t\} \sim IID(o, \sigma^2)$ and is stationary.

Hence, $E[(X_{n+h} - f(X_1, ... X_n))] = 0$ and x_i cannot be used for future values.

Example 2   Simple symmetric random walk
The random walk $\{s_t, t = 0, 1, 2, ...\}$ cumulatively summing iid rvs.

If $s_0 = 0$ and $s_t = X_1 + X_2 + ... X_t, t = 1, 2, ...$

If $\{X_t\}$ is iid and binary $(0, 1)$, then it is called a simple symmetric random walk.

Then, $s_t - s_{t-1} = X_t$ which is stationary IID noise. $E(s_t)$ is non-stationary and ever increasing and $E(s_t^2) = t\sigma^2.$

Example 3   Harmonic regression

   Suppose the time series is having the influence of weather, which can be modelled as periodic wave of fixed known *point*, i.e.

$$X_t = s_t + y_t$$

and $$s_t = a_0 + \sum_{j=1}^{k} (a_{\hat{j}} \cos(\lambda_{\hat{j}} t) + b_j \sin(\lambda_{\hat{j}} t))$$

where $a_{\hat{j}}$ and b_j are unknown parameters and $\lambda_{\hat{j}}$ are fixed frequencies, each being some integer multiple of $2\pi/d$. If n is an integer multiple of d, then frequencies $\lambda_{\hat{j}} = j = 1, 2, ... k$ are integer multiples $2\pi m_1/n, ...,$ $2\pi m_k/n$ of $2\pi/n$.

   We can estimate s_t if $k < 4$ by Fourier analysis and remove it from X_t to obtain Y_t. In any case, the fixed seasonality components are removed for the time series in order to obtain a residual time series which is stationary provided $\{X_t, t = 0, \pm 1, ...\}$ has some statistical property to time-shifted series $\{X_{t+h}, t = 0, \pm 1, ...\}$ for each integer h. We again restrict the attention to the stationarity of second order to let $\{X_t\}$ be a time series with $EX_t^2 < \infty$. The mean function of $\{X_t\}$ is $E(X_t) = \mu_X(t)$ and covariance function

$$\gamma_X(r, s) = Cov(X_r, X_s) = E[(X_r - \mu_X(r))(X_s - \mu_X(s))]$$

for all integers r and s.

   If $\mu_X(t)$ is independent of t, i.e. $\mu_X(t) = \mu_X \ \forall t$ and $r_X(t + h, t)$ is independent of t for each h, then it is weakly stationary.

   The autocovariance function $\gamma_X(h)$ of a weakly stationary series is independent of t for each log h and is equal to $\gamma_X(h, 0) = \gamma_X(t + h, t)$ for any $t = Cov(X_{t+h}, X_t)$.

   The autocorrelation function $\rho_X(h)$ is the normalized autocovariance function and is given by

$$\gamma_X(h)/\gamma_X(0) = Cov(X_{t+h}, X_t).$$

   Due to the linearity property of autocovariance function we have for stationary series X_t, Y_t, Z_t:

$$Cov(aX + bY + C, Z) = a \, Cov(X, Z) + b \, Cov(y, Z)$$

where a, b, c are real-valued constants.

Example 4   White noise

Let $\{X_t\}$ be a sequence of uncorrelated random variables, each with zero mean and variance $\sigma^2 < \infty$. Then, $\{X_t\}$ is stationary and has the same covariance as IID noise. It is denoted as white noise series $\{X_t\} \sim WN(0, \sigma^2)$. Although $IID(0, \sigma^2)$ is $WN(0, \sigma^2)$, the converse is not true.

Example 5 For a first order moving average MA(1) process,

$$X_t = Z_t + \theta Z_{t-1}, t = 0, \pm 1, \pm 2 \ldots.$$

where $\{Z_t\} \sim WN(0, \sigma^2)$ and θ is a real-valued constant.

So, $EX_t = 0$, $EX_t^2 = \sigma^2(1 + \theta^2) < \infty$ and hence MA (1) is always stationary. Its autocovariance function is

$$\gamma_X(t + h, t) = \begin{cases} \sigma^2(1 + \theta^2) & \text{if } h = 0 \\ \sigma^2\theta & \text{if } h = \pm 1 \\ 0 & \text{if } |h| > 1 \end{cases}$$

The autocorrelation function for MA (1) process is, therefore,

$$\rho_X(h) = \begin{cases} 1 & \text{if } h = 0 \\ \theta/(1 + \theta^2) & \text{if } h = \pm 1 \\ 0 & \text{if } h > 1 \end{cases}$$

Example 6 For a stationary first order autoregressive process AR (1) we have $X_t = \phi X_{t-1} + Z_t; t = 0, \pm 1, \pm 2, \ldots$
where $\{Z_t\} \sim WN(0, \sigma^2)$, $|\phi| < 1$ and Z_t uncorrelated with X_s for each $s < t$.

Since $EZ_t = 0$, we have $EX_t = 0$ and its autocovariance function is given by

$$\gamma_X(h) = Cov(X_t, X_{t-h}) = Cov(\phi X_{t-1}, X_{t-h}) + Cov(Z_t, X_{t-h})$$
$$= \phi \gamma_X(h-1) + 0 = \cdots = \phi^h \gamma_x(0).$$

Noting that $\gamma_X(h) = \gamma_X(-h)$ (even function), we get $\rho_X(h)$ $\phi^{|h|}$, $h = 0$, $\pm 1, \ldots$

Since Z_t is uncorrelated with X_{t-1} and using the linearity of covariance, we get

$$Cov(X_t, Z_t) = Cov(\phi X_{t-1}, + Z_t, Z_t)$$
$$= Var(Z_t)$$
$$= \sigma^2$$
$$\gamma_X(0) = Cov(X_t, \phi X_{t-1} + Z_t)$$
$$= \phi \gamma_X(1) + \sigma^2$$
$$= \phi^2 \gamma_X(0) + \sigma^2$$

Therefore, $\gamma_X(0) = \sigma^2/(1 - \phi^2)$ which is finite if $|\phi| < 1$ and the series is stationary, as desired.

Example 7 Sample autocorrelation function

Since the sample of n sequential values of a time series forms one realization of the process $\{X_t\}$, we can compute its sample mean, sample autocovariance function and sample autocorrelation function, which are estimates of population values. The formulae are:

$$\bar{x} = \frac{1}{n} \sum_{i=1}^{n} x_i$$

$$\hat{r}(h) = \sum n - |h|(x_{t+|h|} - \bar{x})(x_t - \bar{x}), -n < h < n.$$

$$\hat{\rho}(h) = \hat{r}(h)/\hat{r}(0), -n < h < n.$$

Usually, we take $h \leq n/4$, so that the estimates are reasonably precise. The divisor n in the above maximum likelihood estimator formulae ensures that the covariance and correlation matrices are non-negative definite, which is essential for subsequent eigenstructure analysis of these matrices.

(Note that unbiased estimators by divisor $(n-1)$ for covariance or correlation should not be used for the above reason.)

1.3 TIME DOMAIN ANALYSIS

Here, the parameter t (= time or spatial distance along a straight line) is of prime interest and all statistical analyses are made with respect to time t. This has the advantage that the operations and forecasting are easily visualized in the domain of parameter t. Assume we have a non-seasonal or deseasonalized time series

$$X_t = m_t + Y_t, t = 1, 2, \dots n \qquad \qquad \dots(16)$$

where $EY_t = 0$.

We can use non-parametric methods such as moving average and spectral smoothing for estimation of trend (signal) m_t, but these will not be useful for a model building process.

Exponential smoothing is causal and useful for forecasting. The smoothed value at present time is the forecast of the next value. For purposes of model building, we either fit a polynomial trend or suitably difference the series (Box and Jenkins method). We then use the residuals of either approach as stationary time series for modelling and forecasting.

1.3.1 Smoothing

A two-sided MA is given by

$$W_t = (2q + 1)^{-1} \sum_{j=-q}^{q} X_{t-j} \qquad \text{...(17)}$$

where q is a non-negative integer. Then, for $q + 1 \le t \le n - q$

$$W_t = (2q + 1)^{-1} \sum_{j=-q}^{q} m_{t-j} + (2q + 1)^{-1} \sum_{j=-q}^{q} Y_{t-j} \cong m_t \qquad \text{...(18)}$$

assuming m_t is approximately linear over interval $[t - q, t + q]$, and average of error terms is nearly zero.

Therefore, $\qquad \hat{m}_t = (2q + 1)^{-1} \sum_{j=-q}^{q} X_{t-j}, \ q + 1 \le t \le n - q \qquad \text{...(19)}$

If $t \le q$ or $t > n - q$, we replace $X_t = X_1$ for $t < 1$ and $X_t = X_n$ for $t > n$ for practical purposes of smoothing.

Also, we can think $\{\hat{m}_t\}$ of Eq. (19) as a process obtained by applying a linear filter (operator) in X_t series, i.e. $\hat{m}_t = \sum_{-\infty}^{\infty} a_j X_{t-j}$ with weights $a_j = (2q + 1)^{-1}, -q \le j \le q$.

This a low-pass filter as it removes the rapidly-fluctuating (high frequency) component $\{\hat{Y}_t\}$ to leave the smoothly-varying estimated trend function $\{\hat{m}_t\}$. If q is large, this filter will attenuate noise but allow linear trend functions $m_t = C_0 + C_1 t$ to pass without any distortion. But m_t may not be linear, so q should be kept a small enough value which allows noise also to pass through the filter. By suitable filter design, i.e. choice of a_j weight, we can attenuate all the noise but allow the third or lower degree polynomial trends to pass through a moving average filter. For example, Spencer 15-pt MA filter passes third degree polynomials without distortion with weights

$$a_j = 0, \ |j| > 7 \quad \text{and} \quad a_j = a_{-j}, \ |j| \le 7$$

and $\qquad [a_0, a_1, ..., a_7] = \dfrac{1}{320} [74, 67, 46, 21, 3, -5, -6, -3]$

For a second order polynomial trend, we need a 5-pt moving average filter.

A two-sided MA filter is non-causal, so we can use a causal (one-sided) MA filter which is also called experimental smoothing. We obtain

$$\hat{m}_t = aX_t + (1-a)m_{t-1}, t = 2, ..., n \qquad ...(20)$$

where $a \in [0, 1]$ and $\hat{m}_1 = X_1$.

This implies for $t \geq 2$, $\hat{m}_t = \sum_{j=0}^{t-2} a(1-a)^j X_{t-j} + (1-a)_j X_1$, with weights decreasing exponentially as $(1-a)^j$.

A third method of smoothing is eliminating the high frequency noise in data by Fourier analysis and retaining the low frequency trend (signal) functions. If $f = 1$, the series remains unchanged; so we use $f << 1$.

1.3.2 Trend estimation for modelling

The simplest method of estimation is to assume that the trend is a low order polynomial function (continuous) and it can be estimated by least square fitting of the random errors, i.e. by minimization of the sum of squares $\sum_{t=1}^{m} (x_t - m_t)^2$, where $m_t = b_0 + b_1 t + b_2 t^2$ (say). Then, the sum of squares of errors is given by $\sum_{t=1}^{n} (x_t - b_0 + b_1 t + b_2 t^2)^2$. Then, differentiating with respect to b_0, b_1 and b_2 and solving the normal equation gives the parameter b_i's of the polynomial trend and variance of error, σ^2.

A second method is to difference this series d times ($d \geq 1$). The differenced series becomes stationary and trend is automatically eliminated. Introducing Box and Jenkins back-shift operator method B such that $Bx_t = x_{t-1}$, we obtain

$$\nabla X_t = X_t - X_{t-1} = (1-B)X_t$$

which eliminates the linear trend. Differencing twice, we get

$$\nabla^2 X_t = (1-B)^2 X_t = X_t - 2X_{t-1} + X_{t-2}$$

This eliminates quadratic trend, and so on. However, over-differencing should be strictly avoided as the residual variance suddenly and greatly increases on over-differencing and then modelling would become useless. Usually, differencing is kept to low order, i.e. $d \leq 2$ so that the trend can be physically and easily interpreted.

For seasonality of period $d (d > 1)$, we can use seasonal differencing operator

$$\nabla_d X_t = X_t - X_{t-d} = (1-B^d)X_t$$

∇_d should not be confused with $\nabla^d = (1 - B)^d$

Once the trend and seasonality inputs are removed from the time series data, the residual series is stationary and we can compute its mean, autocovariance and autocorrelation functions by usual maximum likelihood methods. For large n, each Y_t is iid with finite variable. Hence, the distribution of autocorrelation of an iid sequence is given by $N(0, 1/n)$. Therefore, 95% of sample autocorrelation would lie within $\pm 1.96\sqrt{n}$ or approximately $\pm 2\sqrt{n}$. If $n = 100$, we can compute auto-correlations up to lag25 and thus find out whether Y_t is stationary and normally distributed.

We can also compute a single test statistic (Portmanteau) $Q = n\sum_{j=1}^{n} \hat{\rho}^2(j)$, which has a χ^2 distribution with h degrees of freedom and Q statistic can be tested with $\alpha = 0.05$ level of significance.

A slightly different but more precise Portmanteau test is given by

$$Q = n(n + 2)\sum_{j=1}^{n} \hat{\rho}^2(j)/(n - j)$$

which has χ^2 distribution with h degrees of freedom. We can also use several non-parametric tests such as Sign test, Rank test, Turning point test for checking any trend or wild fluctuations. The residuals should also be tested for normality of distribution by using cumulants, Probit graph, $Q-Q$ plots and the Kolmogorov-Smirnov test. Once the Y_t' series is found to be stationary and normally distributed, then the data can be used for time series modelling and other inferences.

1.4   FREQUENCY DOMAIN ANALYSIS

By frequency domain, we refer to the transformation of a stationary time series $\{X_t\}$ into a form of sinusoidal components with uncorrelated random coefficients. As each of these sinusoids has a characteristic frequency, the representation in the frequency domain is also called spectral analysis and forms a complementary method to visualize the characteristics of the time series using autocovariance function in time domain. Sometimes, the results of spectral analysis would be more illuminating than time-domain methods and especially for study of vector stationary processes and for an analysis of linear filters. In most books, time and frequency domain analyses are thought to be competing but it is better to view these are complementary and hence, both methods provide deeper insights regarding the genesis of time series $\{X_t\}$.

Linear systems

Linear systems are most useful in statistical analysis since superposition of several linear systems remains linear, which enables easier mathematical analysis. For example, a system is linear if and only if the linear components of input yield linear components of output, i.e.

$$\lambda_1 x_1(t) + \lambda_2 x_2(t) = \lambda_1 y_1(t) + \lambda_2 y_2(t)$$

where λ_1, λ_2 and real constants. The linear system $x(t)$ yields output $y(t)$ and is invariant if input-output relationship does not change with time.

Time domain studies

In continuous time (C.T.), $y(t) = \int_{-\infty}^{\infty} h(u) x(t-u) \, du$

and in discrete time (D.T.), $y_t = \sum_{k=-\infty}^{\infty} h_k x_{t-k}$

where $h(u)/h(k)$ are impulse response f_n of linear system.

The system is physically realizable/causal if

$$h(u) = 0, \, u < 0 \quad \text{or} \quad h_k = 0, \, k < 0.$$

The system is stable if the bounded input yields bounded outputs and we deal with stable systems only. A sufficient condition for stability is $\Sigma_k |h_k| < C$ (finite constant).

Some examples

(a) Simple moving average: $y_t = (x_{t-1} + x_t + x_{t+1})/3$

The impulse response function $h_k = \begin{cases} 1/3 & \text{for } k = -1, 0, +1 \\ 0 & \text{otherwise} \end{cases}$

This is not causal but useful as a mathematical smoothing device.

In continuous time, we can have linear differential equation with constant coefficient as

$$T \frac{dy(t)}{dt} + y(t) = x(t)$$

and its equivalent differential equation in discrete time is

$$y_t + \alpha_1 \nabla y_t + \alpha_2 \nabla^2 y_t + \cdots = \beta_0 x_t + \beta_1 \nabla x_t + \beta_2 \nabla^2 x_t + \cdots$$

by successive substitution, we obtain y_t as infinite sum

$$y_t = \sum_{k=-\infty}^{\infty} h_k x_{t-k}$$

with impulse response function $h_k = \begin{cases} (a_1)k, & k = 0, 1, \ldots \\ 0, & k < 0 \end{cases}$

When higher α's and β's are all zero and $\beta_0 = 1$.

(b) Simple delay $y_t = x_{t-d}$

with $h_k = \begin{cases} 1 & k = d \\ 0 & otherwise \end{cases}$

(c) Simple gain $y_t = g x_t$ with impulse response function

$$h_k = \begin{cases} g & k = 0 \\ 0 & otherwise \end{cases}$$

In continuous time, we have $y(t) = x(t - \tau) \, \delta(u - \tau)$ for simple delay and $y(t) = g x(t) \delta(u)$ for simple gain, where $\delta(.)$ is Dirac delta function.

The causal impulse response function is $h(u) = \begin{cases} g \, e^{-(u - \tau)/T} / T, & u > \tau \\ 0, & u < \tau \end{cases}$

in a delayed exponential function with $\tau =$ delay time (constant), $g =$ gain constant and T is a constant which governs the rate of delay in an output.

If input is made at t_0, 1, then output $y(t)$ changes over time t, as shown in Fig. 1.2.

A linear system can be described by the use of impulse response function. Suppose impulse is 1 at $t = 0$ and zero otherwise, then the output in discrete time is given by $y_t = \Sigma h_k x_{t-k} = h_t$. Thus, unit impulse gives unit impulse response function.

Alternatively, the linear system can be described by the step response function

$$S(t) = \int_{-\infty}^{t} h(u) \, du \text{ (continuous time)}$$

or $\quad S(t) = \Sigma_{k < t} h_k$ (discrete time).

For example, in discrete time, suppose input is $x_t = \begin{cases} 0, & t < 0 \\ 1, & t \geq 0 \end{cases}$

then, $y_t = \Sigma_k h_k x_{t-k} = \Sigma_{k \leq t} h_k = S_t$.

So, the output is equal to the step response function and its derivatives in the impulse response function. A step change is easier to provide in a causal system than an impulse at time $t = t_0$. The step response function for a delayed exponential system is given by $S(t) = g[1 - e^{-(t-\tau)/T}], t > \tau$.

Frequency domain studies

Instead of time domain, we can describe the linear system in a frequency domain using the frequency domain function or the transfer function which is the Fourier transform of the impulse response function.

Thus,
$$H(w) = \int_{-\infty}^{\infty} h(u) e^{-iwu} du, 0 < w < \infty \text{ (in C.T.)}$$

$$H(w) = \Sigma_k h_k e^{-ik}, 0 < w < \pi \text{ (in D.T.)}$$

Thus, frequency response functions are equivalent to describe the linear system as autocorrelation and power spectral density functions are equivalent ways for describing the stationary stochastic processes. One function is the Fourier transform of the other function, but we observe that $H(w)$ is much more useful compared to $h(u)$.

Theorem 1

A sinusoidal input to a linear system gives rise—in the steady state—to a sinusoidal output at the *same* frequency. However, the amplitude of the output may change and there may also be a phase shift.

If input $\quad x(t) = e^{iwt} = \cos wt + i \sin wt$, then the output is
$$y(t) = G(w) \{\cos[wt + \phi(w)] + i \sin[wt + \phi(w)]\}$$
$$= G(w)e^{i\phi(w)} x(t)$$

where gain $G(w) = [A^2(w) + B^2(w)]^{1/2}$ and phase shift $\phi(w) = \tan^{-1}[-B(w)/A(w)]$.

Here, $A(w)$ and $B(w)$ are given by

$$A(w) = \int_{-\infty}^{\infty} h(u) \cos wu \, du \quad \text{and} \quad B(w) = \int_{-\infty}^{\infty} h(u) \sin wu \, du.$$

We can easily show that
$$G(w)e^{i\phi(w)} = A(w) - iB(w)$$

$$= \int_{-\infty}^{\infty} h(u) (\cos wu - i \sin wu) du$$

$$= \int\limits_{-\infty}^{\infty} h(u)e^{-iwu}du = H(w).$$

Therefore, in steady state, we have $y(t) = H(w)x(t)$ as defined.

However, this theorem/result is applicable only at a *steady state* (i.e. the sinusoid was applied at $t = \infty$). If sinusoid is applied at $t = 0$, it will take some time to reach a steady state, and the difference between the output for $t = 0$ to $t =$ steady state is called *transient*. The system is stable if the transient tends to zero as $t \to 0$. If the input-output relationship is united as differential (difference) equation, then the steady state solution corresponds to the particular integration of the transient corresponding to complementary function.

It is easier to describe transients through Laplace transform (L.T.) of the impulse response function, since the L.T. is defined for unstable systems as well. So, L.T. method is preferred by engineers, but statisticians prefer to use Fourier transform (F.T.) for stable (linear) systems. However, L.T. should also be used in time series analysis.

The impedance of frequency response function/transfer function $H(w)$ is that to obtain output when inputs are sinusoidals, it is much easier to use $H(w)$ rather than an impulse response function $h(u)$. Thus, when $x(t) = \Sigma_j A_j(w_j) e^{iw_j t}$, we get

$$y(t) = \Sigma_j A_j(w_j) H(w_j)e^{iw_j t}$$

i.e. the complicated convolutes in time domain reduce to multiplication in the frequency domain. So, linear systems are more easily studied in frequency domain.

Although $H(w)$ is defined for $w > 0$, in discrete time, we can compute frequency up to Nyquist frequency π (or $\pi/\nabla t$, where ∇t is the time interval between successive observations). Therefore, linear systems having inputs at higher frequency than π will have a corresponding sinusoid at a frequency in $(0, \pi)$ which gives identical readings at specific intervals of time and gives identical output. This problem is to be kept in mind for analysis of real data.

A necessary and sufficient condition for a linear system to be stable is that the Laplace transform of the impulse response function should have no poles in the right-half plane or on the imaginary axis (Fig. 1.9).

Gain and phase

Frequency response function/transfer function $H(w)$ of a linear system is usually a complex function $H(w) = G(w) e^{i\phi(w)}$, where $G(w) =$ gain, $\phi(w) =$

phase. Although gain $G(w)$ is uniquely defined over $w(0, \pi)$, the phase diagram is not uniquely defined and more difficult to plot.

$$\tan \phi(w) = B(w)/A(w)$$

with $\quad A(w) = \int_{-\infty}^{\infty} h(u) \cos wu \, du \quad$ and $\quad B(w) = \int_{-\infty}^{\infty} h(u) \sin wu \, du$

So, $\quad y(t) = G(w) \cos[wt + \phi(w)]$

If $G(w)$ is positive or negative, then phase is *understood* by an integer multiple of π and often constrained to range $(-\pi/2, \pi/2)$. If $G(w)$ is taken to be positive only, then phase is *understood* by an integer multiple of 2π and is often constrained to $(-\pi, \pi)$. Engineers prefer $\phi(w)$ to be unconstrained with $G(w)$ either positive or negative and use the fact that $\phi(0) = 0$ provided $G(0)$ is finite (Fig. 1.9).

Some examples

(a) Simple MA $y_t = (x_{t-1} + x_t + x_{t+1})/3$ has impulse response function

$$h_k = \begin{cases} 1/3 & k = -1, 0, +1 \\ 0 & otherwise \end{cases}$$

The frequency response function $H(w)$ of this filter is

$$H(w) = \frac{1}{3}e^{-iw} + \frac{1}{3} + \frac{1}{3}e^{iw}$$

$$= \frac{1}{3} + \frac{2}{3}\cos w; \ 0 < w < \pi$$

This is a real function (not a complex function) and so the phase is given by

$$\phi(w) = 0; \ 0 < w < \pi$$

However, $H(w)$ is negative for $w > 2\pi/3$ and so, if we have the correction $G(w)$ to be positive, then

$$G(w) = \left| \frac{1}{3} + \frac{2}{3}\cos w \right|$$

$$= \begin{cases} \dfrac{1}{3} + \dfrac{2}{3}\cos w; & 0 < w < 2\pi/3 \\ -\dfrac{1}{3} - \dfrac{2}{3}\cos w; & 2\pi/3 < w < \pi \end{cases}$$

and $\quad \phi(w) = \begin{cases} 0; & 0 < w < 2\pi/3 \\ \pi; & 2\pi/3 < w < \pi \end{cases}$

The gain is mainly low-pass type and hence simple MA smoothes out local fluctuations (high frequency) and measures the trend (low frequency). In fact, it will be better to allow $G(w)$ to be either positive or negative so that $\phi(w) = 0$ for all w in $(0, \pi)$.

(b) Simple exponential response from a linear system has impulse response function

$$h(u) = ge^{u/T}; \quad u > 0.$$

So, we get $H(w) = g(1 - iwT)/(1 + w^2 t^2); \quad w > 0$

Therefore, $G(w) = g / (1 + w^2 T^2)^{\frac{1}{2}}$ and $\phi(w) = -Tw$

As frequency w increases, $G(w)$ decreases making it a low-pass filter. If we take $\phi(0) = 0$, the phase becomes increasingly negative as w increases until output is out of phase with the input.

(c) *Linear system with phase delay*

$$y(t) = x(t - \tau); \quad \tau = \text{delay constant}$$

$$h(u) = \delta(u - \tau) \quad \text{and} \quad H(w) = \int_{-\infty}^{\infty} \delta(u - \tau)\, e^{-iwu} du = e^{iwt}.$$

The gain $G(w) = 1$ and   phase $\phi(w) = w\,\tau$

More general inputs and outputs

In time domain, we have $y(t) = \int_{-\infty}^{\infty} h(u)x(t - u)du$, which is difficult to evaluate unless $x(t)$ is a simple in form. But in frequency domain, if Fourier transform (F.T.) $Y(w)$ exists, then

$$Y(w) = H(w) \cdot X(w)$$

where $H(w)$ and $X(w)$ are F.T. of $H(u)$ and $x(t)$ and provided they exist. If $x(t)$ is a stationary process, it has a continuous power spectrum and we get Theorem 2.

Theorem 2

Consider a stable linear system with gain function $G(w)$ and assume that the input $X(t)$ is stationary having a continuous power spectrum $f_X(w)$. Then, output $Y(t)$ is also a stationary process where power spectrum $f_Y(w)$ is given by $f_y(w) = G^2(w)f_x(w)$. This holds for discrete time as well.

Proof

Omitted but simple.

We take F.T. of time domain autocovariance function $\gamma_y(\tau)$ to obtain

$$f_y(w) = H(w)H(w)f_x(w) = G^2(w)f_x(w)$$

where $H(w)$ = frequency response function and $G(w) = |H(w)|$ is the gain function of the linear system.

Some examples

(a) MA process $X_t = B_0 z_t + \cdots + B_q z_{t-q}; z_t \sim N(0, \sigma_z^2)$

So, $\qquad H(w) = \sum_{j=0}^{q} B_j e^{-iwj}$ and $f_z(w) = \sigma_z^2 / \pi$

$$f_x(w) = \left| \sum_{j=0}^{q} B_j e^{-iwj} \right|^2 \sigma_z^2 / \pi$$

For MA (1), we have $H(w) = 1 + Be^{-iw}$

and $\qquad G^2(w) = |H(w)|^2 = (1 + B \cos w)^2 + B^2 \sin^2 w$

$$= 1 + 2B \cos w + B^2$$

Therefore, $\qquad f_x(w) = (1 + 2B \cos w + B^2) \sigma_z^2 / \pi.$

(b) AR (1) process $X_t = \alpha X_{t-1}$ or $z_t = X_t - \alpha X_{t-1}$

$$H(w) = 1 - \alpha e^{-iw}; G^2(w) = 1 - 2\alpha \cos w + \alpha^2$$

and $\qquad \sigma_z^2 / \pi = f_z(w) = (1 - 2\alpha \cos w + \alpha^2) f_x(w).$

(c) Differentiation: $Y(t) = dX(t)/dt.$

If input $X_t = e^{iwt}$ (sinusoidal), then output $Y(t) = iwe^{iwt}$ and frequency response function of linear system is given by

$$H(w) = iw; \quad f_y(w) = |iw|^2 f_x(w) = w^2 f_x(w)$$

The system would be stable if variance of the output should be finite. Computing the variance of $Y(t)$, we get

$$Var[Y(t)] = \int_0^\infty f_y(w) \, dw$$

$$= \int_0^\infty w^2 f_x(w) \, dw$$

But we also have $\gamma_x(k) = \int_0^\infty f_x(w) \cos wk \, dw$ and

$$\frac{d^2 \gamma_x(k)}{dk^2} = -\int_0^\infty w^2 f_x(w) \cos wk \, dw$$

So that $Var[Y(t)] = -\left[\dfrac{d^2\gamma_x(k)}{dk^2}\right]_{k=0}$

Thus, $Y(t)$ has finite variance and provided $\gamma_x(k)$ can be differ-entiated twice at $k = 0$, then only we have $f_y(w) = w^2 f_x(w)$.

If two or more linear systems are in series, then the overall frequency response function $H(w)$ is the product of the individual frequency response functions, i.e.

$$H(w) = H_1(w) + H_2(w) + \cdots + H_k(w)$$

and if $H_j(w) = G_j(w)e^{i\phi_j w}$, then $H(w) = \displaystyle\prod_{j=1}^{k} G_j(w)e^{i\left[\Sigma_j \phi_j(w)\right]}$

Then, overall gain is the product of the individual gains and overall phase is the sum of the individual phase shifts. We can design the filters (linear systems) to suit our purpose (i.e. to obtain trends or eliminate trends). A low-pass filter estimates the trend and high-pass filter eliminates it. The difference operator is having a $G(w)$ that is essentially a high-pass filter (i.e. eliminates trends) but does not do so perfectly and cut-off is poor.

We have $y_t = \Sigma_k h_k x_{t-k}$, where y_t is filtered output.

Its frequency response function $H(w) = \Sigma_k h_k e^{-iw_k}$

and $f_y(w) = G^2(w)f_x(w); G(w) = |H(w)|.$

Difference operator $y_t = x_t - x_{t-1}$ and its $H(w) = 1 - e^{-iw}$.

Its gain function $G(w) = [2(1 - \cos w)]^{1/2}$, which is 0 at $w = 0$ and $G(w) = 2$ at $w = \pi$. This is a high-pass filter (trend eliminator) but is does not completely cut off some middle frequencies (i.e. poor cut off on the left side). Therefore, the difference operator is not an excellent trend eliminator, which should be kept in mind in time series modelling (see Fig. 1.8).

Linear system identification

Usually, the spectral form of a linear system could be unknown, so we have to estimate it from the cross-spectra of input with output (identification). Identification is simple if either input is controlled or if the system is not *countered* with noise. In such a case, we apply impulse or step function as input and find the impulse response or step response function as outputs. Alternatively, we can apply sinusoids as inputs at different frequencies and observe the amplitude and phase shifts of the corresponding sinusoidal outputs. This gives us the required corresponding gain and phase diagrams.

But many systems are contaminated with noise $N(t)$. This noise process may not be white noise (i.e. not a purely random process) but is usually assumed to be uncorrelated with input process $X(t)$.

Further difficulty arises when input is observable but not *controllable* (i.e. we cannot make step change to input) leading to a more refined analysis, especially when there is a feedback. These identifications are made in frequency domain and time domain, independently, using cross-spectral analysis input-output (frequency response function); impulse response function (due to Box-Jenkins) of linear systems.

Frequency response function

In presence of noise, we derive cross-spectra of inputs and outputs.

$$Y(t) = \int_0^\infty h(u) X(t-u) du + N(t)$$

subject to $h(u) = 0$ for $u < 0$ (causal/physically relizable)
For convenience, $E(X(t)) = 0$; so, $E(Y(t)) = 0$.
Multiplying by $X(t - \tau)$ and taking expectation, we get

$$\gamma_{xy}(\tau) = \int_0^\infty h(u) \, \gamma_{xx}(\tau - u) du \text{ (Wiener-Hopf equation)}$$

where γ_{xy} is cross-covariance function of $X(t)$, $Y(t)$ and γ_{xx} is auto-covariance function of $X(t)$. If $h(u)$ is known, then the above equation can be solved in principle (convolution integral) but it often easier to solve this in frequency domain.

Taking F.T. of discrete time anlayses (by multiplying $e^{-iw\tau}/\pi$ and solving for $\tau = -\infty$ to $+\infty$), we get

$$f_{xy}(w) = \sum_{\tau=-\infty}^{\infty} \sum_{k=0}^{\infty} h_k e^{-iwk} \gamma_{xx}(\tau - k) e^{-iw(\tau-k)} / \pi$$

$$= \sum_{k=0}^{\infty} h_k e^{-iwk} f_x(w)$$

$$= H(w) f_x(w)$$

Estimates of $f_x(w)$ and $f_{xy}(w)$ can now be used to obtain estimate of $H(w)$ as $\hat{H}(w) = \hat{f}_{xy}(w) / \hat{f}_x(w)$.

We usually write $H(w) = G(w) e^{i\phi(w)}$ and estimate gain and phase separately. We have Eq. $\hat{G}(w) = |\hat{H}(w)| = |\hat{f}_{xy}(w) / \hat{f}_x(w)|$

$$= | \hat{f}_{xy}(w) | / \hat{f}_x(w)$$

since $f_x(w)$ is real $= \hat{\alpha}_{xy}/\hat{f}_x(w)$ (cross amplitude spectra).

Also, $\tan \phi(w) = -\hat{q}(w)/\hat{c}(w)$

where $q(w)$ and $c(w)$ are the quadrature- and co-spectra, respectively. So, even in the presence of $N(t)$ we obtain estimates of gain and phase of the linear system.

In discrete time case,

$$f_y(w) = H(w)\overline{H(w)}f_x(w) + f_n(w)$$

But $H(w)\overline{H(w)} = G^2(w) = c(w)f_y(w)/f_x(w)$

So, $f_n(w) = f_y(w)[\wedge - c(w)]$, for which $f_n(w)$ can be estimated. $c(w)$ is coherency, at frequency w, varying from 0 to 1, and measures linear correlation between input and output at frequency w. In proper estimates the frequency response function of linear systems may be transformed in order to give estimates of impulse response function but this is not recommended as phase-leads/lags rarely provide the desired estimates of the time domain relationship of economic lagged model time series.

Box-Jenkins approach

In this approach, input and output at both differenced time series until stationarity and also mean-corrected, $Y_t = \sum_{k=0}^{\infty} h_k X_{t-k} + N_t$

Multiplying X_{t-m} and taking expectations, we have

$$\gamma_{xy}(m) = h_0\gamma_{xx}(m) + h_1\gamma_{xx}(m-1) + \cdots$$

by assuming N_t is uncorrelated with input. If we assume weights $\{h_k\}$ to be effectively zero beyond $k = K$, the first $K + 1$ have form $= 0, 1, \ldots\ldots K$ can be fitted for $K + 1$ unknown $h_0, h_1, \ldots, h_K$ using estimates of γ_{xy}, γ_{xx}.

But the solutions are usually not very good and also K has to be known. Box-Jenkins propose two modifications to solve this problem. The first modification is prewhitening, which removes basic auto-correlation within the input and output series. A second modification is using an alternative form of lagged equation requiring fewer parameters.

$$Y_t - \delta_1 Y_{t-1} - \cdots - \delta_r Y_{t-r} = w_0 X_{t-b} - w_1 X_{t-b-1} \cdots - w_s X_{t-b-s}$$

where b = delay of system ≥ 0 (integer).

$\delta(B)Y_t = w(B)X_{t-b}$ (transfer function model—a misnomer as 'transfer function' is used in place of transfer function of impulse response function).

Suppose the differenced time series is stationary and Box-Jenkins model is

$$\phi(B)X_t = \theta(B)z_t$$

Then the inputs can be transferred to white noise by $\phi(B)\theta^{-1}(B)X_t = z_t = \alpha_1$ and we apply the same transfer to the output to obtain

$$\phi(B)\theta^{-1}(B)\, Y_t = \beta_t$$

and calculate correlation covariance between fitted input and output $\{\alpha_t\}$ and $\{\beta_t\}$.

This gives better impulse response function

$$h(B) = h_0 + h_1 B + h_2 B^2 + \cdots$$

so that $\qquad Y_t = h(B)X_t + N_t.$

Then

$$
\begin{aligned}
\beta_t &= \phi(B)\theta^{-1}(B)Y_t \\
&= \phi(B)\theta^{-1}(B)\,[h(B)X_t + N_t] \\
&= h(B)\alpha_t + \phi(B)\theta^{-1}(B)N_t
\end{aligned}
$$

and $\qquad \gamma_{\alpha\beta}(m) = h_m Var(\alpha_t)$

since $\{\alpha_t\}$ is purely random and N_t is uncorrelated with $\{\alpha_t\}$. This is much simpler than the time domain approach.

Hence, we obtain $\hat{h}_m = c_{\alpha\beta}(m)/s_\alpha^2$, ratio of sample cross-correlation to sample variance.

These equations can be used for modelling, forecasting as well as control. The time domain and frequency domain approaches are compliments rather than rivals, so we should try both the approaches. Another way to convert non-stationary to stationary series is by passing it through high-pass filter rather than differencing the series (Box-Jenkin's method).

System with feedback

It can be solved (Akaike, 1967) in frequency domain using methodology for closed-loop systems.

Then,

$$f_{xy}/f_n = \left(Hf_v + \bar{J}f_n\right)/\left(f_v + J\bar{J}f_n\right)$$

only if $f_n \equiv 0$ or $J \equiv 0$ is the ratio $f_{xy}/f_x = H$.

Thus, estimate of H obtained by $\hat{f}_{xy}/\hat{f}_n$ will be poor unless f_n/f_v is small. In particular, if $f_v \equiv 0$, $\hat{f}_{xy}/\hat{f}_n$ will provide an estimate of J^{-1} and *not* of H. A similar analysis can be done in time domain.

Hence, for open loop, these closed-loop procedures would not be applicable/appropriate. An approach is to deliberately add $V(t)$ noise with known clusters and analyse. If feedback is present, the cross-correlation and cross-spectral analysis of raw data will be misleading/meaningless.

1.4.1   Spectrum

The spectral density of $\{X_t\}$ is defined as

$$f(\lambda) = \frac{1}{2\pi} \sum_{h=-\infty}^{\infty} e^{-ih\lambda} \, y(h); \quad -\infty < k < \infty$$

which converges absolutely $|e^{ih\lambda}|^2 = \cos^2(h\lambda\alpha) + \sin^2(h\lambda) = 1$.

Since cos and sin have periodicity 2π as that of f, it sufficient to restrict values of f in the interval $(-\pi, +\pi)$.

Some basic properties of spectral density f are $f(\lambda) = f(-\lambda)$.

$$f(\lambda) \geq 0 \text{ for } \lambda \in (-\pi, \pi] \quad \text{and} \quad r(k) = \int_{-\pi}^{\pi} \cos(k\lambda) f\lambda \, d\lambda.$$

The last equation gives $r(k)$ of a stationary time series with absolutely summable acvf as low Fourier coefficients of non-negative even function in $(-\pi, \pi]$.

However, even if $\sum_{h=-\infty}^{\infty} |r(h)| = \infty$, we still define spectral density of stationary time series $\{X_t\}$ if

$$r(h) = \int_{-\pi}^{\pi} e^{ih\lambda} \, f(\lambda) \, d\lambda \text{ for all integer } h \text{ and } f(\lambda) \geq 0 \text{ for all } \lambda \in (0, \pi].$$

A real valued function f on $(0, \pi]$ is spectral density of a stationary process if:

(i) $f(\lambda) = f(-\lambda)$;

(ii) $f(\lambda) \geq 0$;   and

(iii) $\int_{-\pi}^{\pi} f(\lambda) \, d\lambda < \infty$.

An absolutely summable function $r(.)$ can be expressed as Fourier transform of discrete distribution for

$$F(\lambda) = \begin{cases} 0; & \text{if } \lambda < -w \\ \frac{1}{2}; & \text{if } -w_j \leq \lambda < w \\ 1.0; & \text{if } \lambda \geq w_j \end{cases}$$

so that $\quad r(h^2)\cos(wh) = \int(-\pi, \pi] \, e^{ih\lambda} \, dF(\lambda).$

This is called spectral response of autocovariance function and $F(\lambda)$ is referred to as the spectral distribution function (which is not differentiable for this case to obtain spectral density function $f(\lambda)$).

Example: Linear combination of sinusoids

Let $\qquad X = \sum_{j=1}^{k}\left(A_j \cos\left(w_j t\right) + B_j \sin\left(w_j t\right)\right)$

where $0 < w_j < \pi$ for each j and Aj's, B_j's uncorrelated with $E(A_j) = 0$ and $Var\,(A_j) = Var\,(B_j) = \sigma_j^2$.

It has ACVF $r(h) = \sum_{j=1}^{k} \sigma_j^2 \, \cos(w_j h)$

and spectral distribution function $F(\lambda) = \sum_{j=1}^{k} \sigma_j^2 \, F_j(\lambda)$

where $\qquad F_j(\lambda) = \begin{cases} 0; & \text{if } \lambda < -w \\ \frac{1}{2}; & \text{if } -w_j \leq \lambda < w_j \\ 1.0; & \text{if } \lambda \geq w_j \end{cases}$

The sample path of this is an approximate sinusoid with frequency dominant w_j and period $2\pi/w_j$.

Every zero-mean stationary process can be expressed as a superposition of uncorrelated sinusoids with frequencies $w \in [0,\,\pi]$. A stationary process is, therefore, a superposition of infinitely many sinusoids and can be represented as a stochastic integral

$$X_t = \int(-\pi, \pi] e^{ih\lambda} \, dz(\lambda)$$

where $\{z(\lambda), -\pi < \lambda \leq \pi\}$ is a complex valued process with orthogonal increments (uncorrelated increments). This is called spectral representation of the process X_t, which is paralleled to spectral representation of autocovariance function and by representing

$$dz(\lambda) = \begin{cases} (A_j + iB_j)/2; & \text{for } \lambda = -w_j \\ (A_j - iB_j)/2; & \text{for } \lambda = w_j \\ 0; & \text{otherwise} \end{cases}$$

we obtain

$$E\left(dz(\lambda)\,\overline{dz(\lambda)}\right) = \begin{cases} \sigma_j^2/2 & \text{for } \lambda = \pm w_j \\ 0 & \text{otherwise} \end{cases}$$

which is equal to jump in spectral distribution function. $F(\lambda) - F(\lambda-)$, for a discrete spectrum or $f(\lambda)\,d\lambda$ for a continuous spectrum. Therefore either a large jump in discrete spectrum distribution function or a large peak in spectral density at frequency $\pm w$ indicates strong sinusoidal components at w. The period of sinusoid with frequency w radians per unit time is $2\pi/w$.

1.4.2 Time invariant Linear Filters (TLF)

Earlier, we have shown that these filters are useful for smoothing, estimating the trend, eliminating the seasonal and/or trend components of the data. A linear process is output of a TLF applied to a white noise input series, i.e.

$$Y_t = \sum_{k=-\infty}^{\infty} c_{t,k}\, X_t; \quad t = 0, \pm 1, \dots.$$

The filter $c_{t,k}$ is time-invariant if $c_{t,t-k} = \psi_k$ (independent of t).
In this case,

$$Y_t = \sum_{k=-\infty}^{\infty} \psi_k\, X_{t-k}; \quad t = 0, \pm 1,\dots$$

and therefore, $\quad Y_{t-s} = \sum_{k=-\infty}^{\infty} \psi_k\, X_{t-s-k}; \quad t = 0, \pm 1,\dots$

so that a time-shifted process is obtained by the repeated use of the same linear filter $\psi = \{\psi_j, j = 0, \pm 1, \dots\}$. TLF ψ is said to be causal if $\psi_j = 0$ for $j < 0$, as in that case, Y_t is expressible in terms of X_s, $s \le t$. The filter $Y_t = aX_{-t}$; $t = 0, \pm 1, \dots$ is linear but not time invariant since $c_{t,t-k} = 0$ everywhere except for $k = 2t$ when its value is a.

Spectral methods are very useful in describing the behaviour of TLF as well as for filter design in order to suppress high frequency (noise) components in the data. Spectral density of the output of a TLF is closely related to the spectral density of the input series (Figs. 1.7 and 1.8).

If $\{X_t\}$ be a stationary time series with zero mean and spectral density $f_x(\lambda)$ and $\psi = \{\psi_j, j = 0, \pm 1,...\}$ is an absolutely summable TLF (i.e. $\sum_{j=-\infty}^{\infty} |\psi_j| < \infty$), then $Y_t = \sum_{j=-\infty}^{\infty} \psi_j X_{t-j}$; $t = 0, \pm 1,...$ is stationary with mean zero and spectral density

$$f_Y(\lambda) = |\psi(e^{-i\lambda})|^2 f_X(\lambda)$$
$$= \psi(e^{-i\lambda})\,\psi(e^{i\lambda})\,f_X(\lambda)$$

where $\psi(e^{-i\lambda}) = \sum_{j=-\infty}^{\infty} \psi_j\,e^{-ij\lambda}$ is the transfer function of the filter. The squared modulus $|\psi(e^{-i\lambda})|^2$ is called power transfer function of the filter.

Suppose we apply two absolutely summable TLFs ψ_1 and ψ_2 to input series X_t. Then the net effect for the spectral density of the output process $Y_t = \psi_1(B)\psi_2(B)X_t$ is

$$\left|\psi_1\left(e^{-i\lambda}\right)\psi_2\left(e^{-i\lambda}\right)\right|^2 f_X(\lambda).$$

We can remove the seasonal component of period s by differencing out lags (i.e. using ∇_s operator) which has a transfer function $1 - e^{-is\lambda}$. This transfer function is zero at all frequencies that are integer multiples of $2\pi/s$ radians per unit time, so it can remove all the components with periods.

1.5 SPECTRAL ANALYSIS

Introduction

We can represnt a stationary time series as the sum data of sinusoidal outputs with uncorrelated random coefficients. Therefore, the autocovariance function of time series can be similarly decomposed into sinusoids having different frequencies (spectral representation). This is similar to a Fourier representation of a deterministic function and is called frequency domain or spectral analysis of time series. Spectral analysis may be more illuminating (than time domain) in cases where loading forces do not have resonant frequency as the structure. Spectral analysis is also useful for analyses of multivariate stationary processes and for linear filters as computations are compressed by using complex variables.

Let $\{X_t\}$ be a zero-mean stationary time series with ACVF $r(.)$ satisfying $\sum_{h=-\infty}^{\infty} |r(h)| < \infty$. The spectral density of $f(.)$ defined by:

$$f(\lambda) = \frac{1}{2\pi} \sum_{h=-\infty}^{\infty} e^{-ih\lambda} r(h); \quad -\infty < \lambda < \infty$$

where $e^{i\lambda} = \cos \lambda + i \sin \lambda$ and $i = \sqrt{-1}$.

The absolute summability of $r(h)$ indicates that $f(\lambda)$ is convergent since

$$|e^{ih\lambda}|^2 = \cos^2(h\lambda) + \sin^2(h\lambda) = 1.$$

Since sin and cos functions as well as f have period 2π, we confine our attention to values of f in interval $(-\pi, \pi]$. we note that f is even, $f(\lambda) \geq 0$ for all λ and

$$r(h) = \int_{-\pi}^{\pi} e^{ih\lambda} f(\lambda) d\lambda$$

$$= \int_{-\pi}^{\pi} \cos(k\lambda) f(\lambda) d\lambda$$

However, even if $\sum_{h=-\infty}^{\infty} |r(h)| = \infty$, the spectral distributions exist and are unique for all integer values h and $r(h) = \int_{-\pi}^{\pi} e^{ih\lambda} f(\lambda) d\lambda$.

An absolutely summable function $r(.)$ is the ACVF of a stationary time series if it is even and

$$f(\lambda) = \frac{1}{2\pi} \sum_{h=-\infty}^{\infty} e^{-ih\lambda} r(h) \geq 0 \text{ for all } \lambda \in (-\pi, \pi]$$

in which case $f(.)$ is the spectral density of $r(.)$. This is easier to check than non-negative definiteness of ACVF.

Not all ACVF have a spectral density as the spectral distribution function may be discrete, for example

$$X_t = A \cos(wt) + B \sin(wt)$$

where A and B are uncorrelated random variables with mean zero and variance 1 with ACVF $r(h) = \cos(wh)$ with

$$F(\lambda) = \begin{cases} 0 & \text{if } \lambda < -w \\ \frac{1}{2} & \text{if } -w \leq \lambda < w \\ 1 & \text{if } \lambda \geq w \end{cases}$$

We can add uncorrelated sinusoids as

$$X_t = \sum_{j=1}^{k} (A_j \cos(wt) + B_t \sin(wt)), \, 0 < w_1 < ..., w_k < \pi$$

with $\quad E(A_j) = 0$ and $Var(A_j) = Var(B_j) = \sigma_j^2, \, j = 1, ..., k.$

We have ACVF $r(h) = \sum_{j=1}^{k} \sigma_j^2 \cos(w_j h)$ and spectral distribution function

$$F(\lambda) = \sum_{j=1}^{k} \sigma_j^2 F_j(\lambda)$$

where $\quad F_j(\lambda) = \begin{cases} 0 & \text{if } \lambda < -w_j \\ \frac{1}{2} & \text{if } -w_j \leq \lambda < w_j \\ 1 & \text{if } \lambda \geq w_j \end{cases}$

A stationary process is a superposition of infinitely numerous sinusoids and then using the stochastic integral, we have

$$X_t = \int_{(-\pi, \pi]}^{r} e^{ih\lambda} \, dz(\lambda)$$

where $\{z(\lambda), -\pi < \lambda < \pi\}$ is a complex-valued process with orthogonal (uncorrelated) increments. In the simple case of finite linear superposition of sinusoids, we have

$$dz(\lambda) = \begin{cases} (A_j + B_j)/2 & \text{if } \lambda = -w_j; \quad j \in \{1, ...k\} \\ (A_j - iB_j)/2 & \text{if } \lambda = w_j; \quad j \in \{1, ...k\} \end{cases}$$

Then, $\quad E\left(dz(\lambda) \, \overline{dz(\lambda)}\right) = \begin{cases} \sigma_j^2; & \text{if } \lambda = \pm w \\ 0 & \text{otherwise} \end{cases}$

In general, for discrete and continuous spectra, we have

$$E\left(dz(\lambda) \, \overline{dz(\lambda)}\right) \text{eq} = \begin{cases} F(\lambda) - F(\lambda-) & \text{for discrete spectra} \\ f(\lambda)d\lambda & \text{for continuous spectra} \end{cases}$$

This shows that a large jump or a large peak in spectral density at frequency $\pm w$ indicates stationary sinusoid and at w (radians) having a period of $2\pi/w$.

White noise

$\{X_t\} \sim WN(0, \sigma^2)$

Then, $r(0) = \sigma^2$ and $r(h) = 0$ for all $|h| > 0$. The spectral density is uniform (rectangular) as $f(\lambda) = \sigma^2/2\pi, -\pi \leq \lambda \leq \pi$.

A process with such a spectra is called white noise, since each frequency in the spectra contributes equally to the variance of the process.

AR(1): $X_t = \phi X_{t-1} + z_t, z_t \sim WN(0, \sigma^2)$

$$f(\lambda) = \frac{\sigma^2}{\pi(\pi - \phi^2)}\left(1 + \sum_{h=1}^{\infty} \phi^h\left(e^{-ih\lambda} + e^{ih\lambda}\right)\right)$$

$$= \frac{\sigma^2}{2\pi}(1 - 2\phi\cos\lambda + \phi^2)^{-1}$$

if ϕ is positive and of high value $< 1, f(\lambda)$ will be high at low frequencies as $\rho(1)$ is high and positive. If ϕ is negative and high absolute value $|\phi| < 1$, $f(\lambda)$ will be high as λ is high frequency (since $\rho(1)$ negative and ACF is highly fluctuating near low frequencies).

MA(1): $X_t = z_t + \theta z_{t-1}, z_t \sim WN(0, \sigma^2)$

$$f(\lambda) = \frac{\sigma^2}{2\pi}(1 + \theta^2 + \theta(e^{-i\lambda} + e^{i\lambda}))$$

$$= \frac{\sigma^2}{2\pi}(1 + 2\theta\cos\lambda + \theta^2)$$

So, $f(\lambda)$ is high at low frequencies if θ high plus values, it is high frequencies if $|\theta|$ is high negative value. This interpretation is similar to that in case of AR(1) process (see Fig. 1.10).

Periodogram

The periodogram ln(.) of observations can be regarded as sample values of $2\pi f(.)$, where $f(.)$ is spectral density function. The vector $\underline{x} = [x_1, ...x_n]$ can be expressed in complex space C^n as

$$\underline{x} = \sum_{k=-[(n-1)/2]}^{[n/2]} a_k e_k$$

where e_k are orthonormal basis vectors and coefficients $a_k = e_k{}^* x = \dfrac{1}{\sqrt{n}} \sum_{t=1}^{n} x_t e^{-itw_k}$.

The sequence $\{a_k\}$ is called the discrete Fourier transform of the sequence $\{x_1, \dots x_n\}$. The tth component $x_t = \sum_{k=[(n-1)/2]}^{[n/2]} a_k \,(\cos(w_k t) - i\sin(w_k t))$; $t = 1, \dots, n$ showing x_t in linear combination of nine waves as desired.

The periodogram $\quad \ln(\lambda) = \dfrac{1}{n}\left| \sum_{t=1}^{n} x_t\, e^{-it\lambda} \right|^2$.

The squared length of $\underset{=}{x}$ is

$$\sum_{t=1}^{n} |x_t|^2 = \underset{=}{x}{}^* \underset{=}{x} = \sum_{k=[(n-1)]}^{[n/2]} |a_k|^2$$

$$= \sum_{k=[(n-1)2]}^{[n/2]} \ln(wx)$$

If $\sum_{h=-\infty}^{\infty} |r(h)| < \infty$, then

$$2\pi f(\lambda) = \sum_{h=-\infty}^{\infty} r(h)e^{-th\lambda}; \lambda \in (-\pi, \pi]$$

We can estimate $f(\lambda)$ as $\ln(\lambda)/2\pi$. For large n, the periodogram ordinates $(\ln(\lambda_1)........\ln(\lambda_m))$ are approximately distributed as independent exponential r.v.s. with means $2\pi f(\lambda_1)........2\pi f(\lambda_m)$, respectively. But for each fixed λ and $\varepsilon > 0$

$$P[\,|\ln(\lambda) - 2\pi f(\lambda)\,| > \varepsilon] \to p > 0,\ \text{as}\ n \to \infty.$$

Thus, $\ln(\lambda)$ is not a consistent estimator of $2\pi f(\lambda)$. We can get a consistent estimator by bringing out the average of the periodogram estimates over a small interval containing λ in such a manner that frequency interval is of the order of $4\pi/\sqrt{n}$, while the number of periodograms averaged is of the order $2\sqrt{n}$.

Example — MA 1

Simple $\quad$ MA $= Y_t = (2q + 1)^{-1} \sum_{|j| \le}^{\infty} X_{t-j}$

We use $m_n = \sqrt{n}$ and weighting function $w_n(j) = (2m_n + 1)^{-1},\ |j| \le m_n$.

Thus, we obtain

$$\left(2\sqrt{n}+1\right)Var\left(\hat{f}(\lambda)\right) \; \rightarrow \; \begin{cases} 2f^2(\lambda) & \text{if } \lambda = 0 \text{ or } \pi \\ f^2(\lambda) & \text{if } 0 < \lambda < \pi \end{cases}$$

In practice n is fixed, so we have to choose m_n and weight function $\{w(.)\}$ to obtain an unbiased and a more or less precise $f(\lambda)$. We can also use non-parametric methods for spectral estimation (not discussed here).

Time-Invariant Linear Filters

We can use time-invariant linear filters to smooth the data, estimate the trend, eliminate seasonal and/or trend components, etc. A linear process is the output of a time-invariant linear filter (TLF) applied to a white noise input either series or generally,

$$Y_t \; = \; \sum_{k=-\infty}^{\infty} c_{t,k} X_k \, ; \, t = 0, \pm 1, \dots$$

The filter $c_{t,k}$ is said to be time-invariant if the weights $c_{t,t-k}$ are independent of t, i.e. $c_{t,t-k} = \psi_k$.

Thus, $\qquad Y_t \; = \; \sum_{k=-\infty}^{\infty} \psi_k X_{t-k} \quad \text{and} \quad Y_{t-k} = \sum_{k=-\infty}^{\infty} \psi_k X_{t-s-k}$,

so that the time-shifted process $\{Y_{t-s}: t = 0, \pm 1, \dots\}$. The TLF ψ is causal if $\psi = 0$ for $j < 0$, since Y_t is then expressible in terms of $X_s ; s \le t$.

The filter $Y_t = aX_{-t}; t = 0, \pm 1, \dots$ is linear but not time-invariant since $c_{t,t-k} = 0$ except when $2t = k$. Thus, $c_{t,t-k}$ depends on the value of t.

For simple MA

$$Y_t = (2q + 1)^{-1}\sum_{j \le q} X_{t-j} \text{ is a TLF with } \psi_j = (2q + 1)^{-1}; j = -q, \dots, q;$$

$\psi_j = 0$ elsewhere.

Spectral methods are useful for describing the behaviour of time-invariant linear filters and for filter design to suppress high frequency (noise) components. The spectral density of output of a TLF is related to spectral density of input and this is a fundamental relation.

Let $\{X_t\}$ be a time series with mean zero and spectral density $f_x(\lambda)$ and $\psi = \{\psi_j, j = 0, \pm 1, \dots\}$ be an absolutely summable TLF (i.e. $\sum_{j=-\infty}^{\infty} |\psi_j| < \infty$).

Then the time series $Y_t = \sum_{j=-\infty}^{\infty} \psi_j X_{t-j}$ is stationary with mean zero and spectral density

$$f_Y(\lambda) = |\psi(e^{-i\lambda})|^2 f_X f(\lambda)$$
$$= \psi(e^{-i\lambda})\psi(e^{i\lambda})f_X(\lambda)$$

where $\psi(e^{-i\lambda}) = \sum_{j=-\infty}^{\infty} \psi_j e^{-ij\lambda} \cdot \psi(e^{-i\lambda})$ is called the transfer function of the filter and $|\psi(e^{-i\lambda})|^2$ is called the power transfer function of the filter.

Let $\{Y_t\}$ be a stationary time series with mean zero and ACVF

$$\gamma_Y(h) = \sum_{j,k=-\infty}^{\infty} \psi_j \psi_k \gamma_X(h+k-j)$$

Since $\{X_t\}$ has a spectral density $f_X(\lambda)$, we have

$$\gamma_X(h+k-j) = \int_{-\pi}^{\pi} e^{i(h-j-k)\lambda} f_X(\lambda)d\lambda$$

Substituting $\gamma_X(h+k-j)$ in the above equation, we get

$$\gamma_Y(h) = \sum_{j,k=-\infty}^{\infty} \psi_j \psi_k \int_{-\pi}^{\pi} e^{t(h-j+k)\lambda} f_X(\lambda)d\lambda$$

$$= \int_{-\pi}^{\pi} \left(\sum_{j=-\infty}^{\infty} \psi_j e^{-ij\lambda}\right)\left(\sum_{k=-\infty}^{\infty} \psi_j e^{-ik\lambda}\right) e^{ih\lambda} f_X(\lambda)d\lambda$$

$$= \int_{-\pi}^{\pi} e^{ih\lambda} \left|\sum_{j=-\infty}^{\infty} \psi_j e^{-ij\lambda}\right|^2 f_X(\lambda)d\lambda$$

So, the spectral density function of $\{Y_t\}$ is given by
$$f_Y(\lambda) = |\psi(e^{-i\lambda})^2 f_X(\lambda)$$
$$= \psi(e^{-i\lambda})(\psi e^{i\lambda})f_X(\lambda)$$

The spectral density of two sequentially-applied absolutely summable TLFs ψ_1 and ψ_2 has a transfer function $\psi_1(e^{-i\lambda})\psi_2(e^{-i\lambda})$ and spectral density of output $\psi_1(B)\psi_2(B)X_t$ is given by $|\psi_1(e^{-i\lambda})\psi_2(e^{-i\lambda})|^2 f_X(\lambda)$.

A simple MA has a transfer function

$$\psi(e^{-i\lambda}) = D_q(\lambda) = (2q+1)^{-1}\sum_{|j|\leq q} e^{-ij\lambda}$$

$$= \begin{cases} \dfrac{\sin(q+1/2)\lambda}{(2q+1)\sin(\lambda/2)} & \text{if } \lambda \neq 0 \\ 1 & \text{if } \lambda = 0 \end{cases}$$

This is a low-pass filter, which has large value for transfer function near zero and tapers off to zero for larger frequencies. An ideal low-pass filter is given by

$$\psi(e^{-i\lambda}) = \begin{cases} 1 & \text{if } |\lambda| \leq w_c \\ 0 & \text{if } |\lambda| > w_c \end{cases}$$

where w_c is prior cut-off value. The corresponding linear filter is, expanding $\psi(e^{-i\lambda})$ as a Fourier series:

$$\psi(e^{-i\lambda}) = \sum_{j=-\infty}^{\infty} \psi_j e^{-ij\lambda}$$

with coefficients $\psi_j = \dfrac{1}{2\pi} \displaystyle\int_{w_c}^{-w_c} e^{ij\lambda} \, d\lambda = \begin{cases} (w_c/\pi) & \text{if } j = 0 \\ \dfrac{\sin(jw_c)}{j\pi} & \text{if } |j| > 0 \end{cases}$

Usually Fourier series may be truncated at suitable value if $j = q$ for a low-pass filter ($q < 10$).

Spectral density of ARMA (p, q)

Let $\{X_t\}$ is a causal ARMA (p, q) process satisfying $\phi(B)X_t = \theta(B)z_t$, then

$$f_X(\lambda) = \frac{\sigma^2 \left| \theta(e^{-i\lambda}) \right|^2}{2\pi \left| \phi(e^{-i\lambda}) \right|^2}, \, -\pi \leq \lambda \leq \pi$$

This is ratio of two trigonometric polynomials, so it is called a rational spectral density.

For ARMA (1,1) $f_X(\lambda) = \dfrac{\sigma^2(1 + \theta^2 + 2\theta \cos \lambda)}{2\pi(1 + \phi^2 + 2\phi \cos \lambda)}.$

Spectral density of AR(2) process is given by

$$f_X(\lambda) = \sigma^2/2\pi(1 - \phi_1 e^{-i\lambda} - \phi_2 e^{-2i\lambda})(1 - \phi_1 e^{i\lambda} - \phi_2 e^{2i\lambda})$$

$$= \sigma^2/2\pi(1 + \phi_1^2 + 2\phi_2 + \phi_2^2 + 2(\phi_1\phi_2 - \phi_1) \cos \lambda - 4\phi_2 \cos^2 \lambda)$$

Differentiating the denominator w.r.t. $\cos \lambda$ and equating to zero, we obtain

$$\cos \lambda = \frac{\phi_1 \phi_2 - \phi_1}{4\phi_2}$$

which gives the peak of spectral density.

Alternatively, we can fit ARMA model to data by additionally using AICC criterion and then compute spectral density to the fitted ARMA model.

Time series exhibiting regular and repetitive behaviour over time is of fundamental importance since this shows that adjacent data are no longer completely independent over time as assumed in classical statistics. It is natural to predict the present value by regressing on the by past values through ARIMA or state-space models. An alternative may be to decompose time series into its regular components, which are expressed as periodic variations represented through Fourier frequencies being driven by sines and cosines to produce linear combinations of sine and cosine functions.

Transform of time to frequency domain is to:

(i) arrange the narration in terms of principal components variances (power spectrum) of these periodic combinations; and

(ii) evaluate each periodic (uncorrelated) contribution as likelihood of independent random variables.

We use a time-invariant linear filter to correlate time domain and frequency domain, approaches. Coherence (special linear transformation) measures the performance of best linear filters relating the two series. Linear filter also isolates a signal embedded in a noise.

Since sine waves oscillate over a period of length 2π. So,

$x_t = A \sin(2\pi f t + \phi)$ or

$x_t = u_1 \sin(2\pi f t) + u_2 \cos(2\pi f t)$ (more useful in time series analysis)

$u_1 = A \cos \phi$ and $u_2 = A \sin \phi$ are transformed independent r.v.s.

$A = (u_1^2 + u_2^2)^{1/2}$; $\phi = \tan^{-1}(u_2/u_1)$.

The equation is generalized by

$$x_t = \sum_{k=1}^{q} (u_k \sin 2\pi f_k t) + (u_{2k} \cos 2\pi f_k t)$$

with $v(u_k) = \sigma_k^2$ and $f_k = 1, 2, \ldots$ to model superposed process.

$$r(h) = \sum_{k=1}^{q} \sigma_k^2 \cos(2\pi f_k h)$$

Example

Simple MA $x_t = \dfrac{1}{3}(z_{t-1} + z_t + z_{t+1})$

$$\gamma_X(h) = \frac{\sigma^2}{9}(3 - |h|) \text{ for } |h| \leq 2 \quad \text{and} \quad 0 \text{ for } |h| > 2$$

$$f_X(f) = \sum_{1/2}^{-1/2} \exp\{-2\pi fh\}. \gamma_X(h); \quad h = 0, \pm 1, \pm 2, \ldots$$

$$= \frac{\sigma^2}{9}[3 + 4\cos(2\pi f) + 2\cos(4\pi f)]$$

Power spectrum and cross spectrum

Stationary time series is considered to have periodic components in proportion to their variances and this is fundamental in spatial representation of the autocovariance function. For weekly stationary by a process, its autocovariance $\gamma(h)$ at $|\log h|$ can be decomposed into monotonically-decreasing function $F(v)$, the spectral distribution function which is held for $-1/2 \leq v \leq 1/2$ and $f(0) = r(0)$ so that

$$\gamma(h) = \int_{-1/2}^{1/2} e^{2\pi ivh} \, dF(v)$$

This is exactly similar to the characteristic function for probability density function of random variables. A periodic stationary random process with frequency v_0 can be represented as

$$x_t = u_1 \sin(2\pi v_0 t) + u_2 \cos(2\pi v_0 t)$$

where u_1 and u_2 are independent zero mean r.v.s. with equal variance σ^2.

One cycle is completed at time period $1/v_0$ and the process makes exactly v_0 cycles per unit, for $t = 0, \pm 1, \pm 2, \ldots$ and wavelength $\lambda_0 = 2\pi v_0$. Therefore,

$$\gamma_X(h) = \sigma^2 \cos(2\pi v_0 h)$$

$$= \frac{\sigma^2}{2} e^{-2\pi i v_0 h} + \frac{\sigma^2}{2} e^{2\pi i v_0 h}$$

$$= \int_{-1/2}^{1/2} e^{2\pi i vh} dF_x(v)$$

where $F_X(v) = \begin{cases} 0 & v < -v_0 \\ \sigma^2/2 & -v_0 \le v < v_0 \\ \sigma^2 & v \ge v_0 \end{cases}$

If $\gamma(h)$ is absolutely summable over h, we can convert the relationship to obtain

$$f(v) = \sum_{h=-\infty}^{\infty} r(h)_t e^{-2\pi i v h} ; \quad -1/2 < v < 1/2$$

This helps in the study of discrete stationary time series. A Fourier transformation pair exists and the following relationship is unique

$$A(v) = \sum_{t=-\infty}^{\infty} a_t e^{-2\pi i vt} \quad \text{and} \quad a_t = \int_{-1/2}^{1/2} A(v) e^{2\pi i vt}$$

where a_t is a generating function which is absolutely summable over t. The variance of process x_t is represented as

$$\text{Var}(x_t) = \gamma(0) = \int_{-1/2}^{1/2} f(v) dv$$

which is the integrated spectral density over all frequencies v.

However, a linear transformation (filter) can isolate the variance in certain frequency interval or bands. This is similar to an analysis of variance (ANOVA) where columns or block effects are the frequencies v.

Example

Let z_t be iid random variable with $t = 1$ to 256 with zero means and variance σ^2.

Its autocovariance is given by

$$\gamma(h) = \sigma^2 \text{ for } h = 0 \quad \text{and} \quad \gamma(h) = 0 \text{ otherwise.}$$

The power spectrum is $f_z(v) = \sigma^2$ for $-1/2 < v < 1/2$ with equal power at all v.

Therefore, white noise spectrum is uniform over $(-1/2, 1/2)$ and $\sigma^2 = 1$.

The spectrum for a three-point moving average series:

$$x_t = \frac{1}{3}(z_{t-1} + z_t + z_{t+1})$$

can be computed as $\gamma_x(h) = \dfrac{\sigma^2}{9}(3 - |h|)$ for $|h| \le 2$ and $\gamma_x(h) = 0$ for $|h| > 2$.

So,

$$f_x(\nu) = \sum_{h=-2}^{2} \gamma_x(h) \exp\{-2\pi \nu h\}$$

$$= \frac{\sigma^2}{9}\left\{e^{-4\pi i\nu} + e^{4\pi i\nu}\right\} + \frac{2\sigma^2}{9}\left\{e^{-2\pi i\nu} + e^{2\pi i\nu}\right\} + \frac{3\sigma^2}{9}$$

$$= \frac{\sigma^2}{9}[3 + 4\cos(2\pi\nu) + 2\cos(4\pi\nu)]$$

Therefore, $f_x(0) = \sigma^2$, $f_x(1/4) = \sigma^2/.9$ and $f_x(1/2) = \sigma^2/9$ which gives minimum power $f(\nu) = 0$ at $\nu = .34$.

Thus, lower frequencies ($\nu < 0.2$) have greater power than higher (faster) frequencies (say $\nu > 0.2$).

For an AR (2) having $x_t - \phi_1 x_{t-1} - \phi_2 x_{t-2} = z_t$ with $\phi_1 = 1.0$, $\phi_2 = -.9$ and $\sigma^2 = 1$, we have strong periodicity making a cycle for every 5 points.

$$\gamma_z(h) = 2.81\gamma_x(h) - 1.90\,[\gamma_x(h+1) + \gamma_x(h-1)] + .90\,[\gamma_x(h+2) + \gamma_x(h-2)]$$

or $\gamma_z(h) = \displaystyle\int_{-1/2}^{1/2} [2.81 - 3.80\cos(2\pi\nu) + 1.80\cos(h\pi\nu)]e^{2\pi h i\nu} f_x(\nu)d\nu.$

By Fourier transform $f_x(\nu) = \sigma^2/[2.81 - 3.80\cos(2\pi\nu) + 1.80\cos(h\pi\nu)]$.

This gives strong power concentrated at $\nu = 1.6$ per point or a period of six and seven cycles per point and very little power at other frequency. So, filtering white noise through an AR(2) operator gives a concentrated power (variance) in a very narrow band.

We can extend this relation to several jointly stationary series x_t and y_t, where coherence at any frequency ν is similar to the component of linear correlation between two r.v.s. X and Y in classical statistics. The covariance function for x_t and y_t is given by

$$\gamma_{xy}(h) = E[(x_{t+h} - \mu_x)(y_t - \mu_y)]$$

which has representation

$$\gamma_{xy}(h) = \int_{-1/2}^{1/2} f_{xy}(\nu)e^{2\pi i\nu h}d\nu, \; h = 0, \pm 1, \pm 2$$

and cross-spectrum is defined by Fourier transform

$$f_{xy}(v) = \sum_{h=-\infty}^{\infty} \gamma_{xy}(h)e^{-2\pi i vh}, -1/2 \le v \le 1/2$$

or $\quad f_{xy}(v) = c_{xy}(v) - iq_{xy}(v)$

where the co-spectra is $\qquad c_{xy}(v) = \sum_{h=-\infty}^{\infty} \gamma_{xy}(h)\cos(2\pi vh)$

and quadrature spectrum is $\quad q_{xy}(v) = \sum_{h=-\infty}^{\infty} \gamma_{xy}(h)\cos(2\pi vh)$

Since $\gamma_{xy}(h) = \gamma_{xy}(-h)$, we have $f_{yx}(v) = \overline{f_{xy}(v)}$

where complex conjugate is written by overbar symbol.

This implies $c_{yx}(v) = c_{yx}(v)$ even function and $q_{yx}(v) = -q_{xy}(v)$ (odd function).

The squared coherence function at frequency v is

$$\rho_{y.x}^2(v) = |f_{yx}(n)|^2/f_x(v) \cdot f_y(v)$$

which is analogous to squared multiple correlation coefficients

$$\sigma_{yx}^2 = \sigma_{yx}^2/\sigma_x^2\sigma_y^2$$

We can find the cross-spectra and coherence of a process with a three-point MA process for demonstration.

Let x_t be a stationary process with spectral density $f_x(v)$ and a three-point MA process is

$$y_t = \frac{1}{3}(x_{t-1} + x_t + x_{t+1})$$

Then,

$$\gamma_{xy}(h) = \frac{1}{3}(\gamma_x(h+1) + \gamma_x(h) + \gamma_x(h-1))$$

$$= \frac{1}{3}\int_{-1/2}^{1/2} [1 + 2\cos(2\pi v)]f_x(v)e^{2\pi i vh}dv$$

The unique Fourier transform for spectral representation is given by

$$f_{xy}(v) = \frac{1}{3}[1 + 2\cos(2\pi v)]f_x(v)$$

so that cross-spectra is real in this case. The spectral density of y_t is

$$f_y(v) = \frac{1}{9}[3 + 4\cos(2\pi v) + 2\cos(4\pi v)]f_x(v)$$

$$= \frac{1}{9}[1 + 2\cos(2\pi v)]^2 f_x(v)$$

The squared coherence $\rho^2_{y.x}(v) = 1$ for all values of v.

For vector stationary process, similarly we have autocovariances maximized at

$$\Gamma(h) = E\left[\left(x_{t+h} - \mu\right)\left(x_t - \mu\right)'\right]$$

satisfying $\sum_{h=-\infty}^{\infty} |\gamma_{jk}(h)| < \infty$ for all $j, k = 1$ and has the representation

$$\Gamma(h) = \int_{-1/2}^{1/2} e^{2\pi i v h} f(v)\, dv,\, h = 0, \pm 1, \pm 2 \ldots$$

as the inverse transform of spectral density matrix

$$f(v) = \{f_{jk}(v)\};\quad j, k = 1, \ldots, p$$

with elements equal to cross-spectral couples. The matrix $f(v)$ has representation

$$f(v) = \sum_{h=-\infty}^{\infty} \Gamma(h)e^{-2\pi i v h};\quad -1/2 < v < 1/2$$

Linear filters

Let input series be x_t and output series be y_t with $t = 0, \pm 1, \pm 2$ and the linear coefficients designated as a_t. The linear relation of output to input is of the form

$$y_t = \sum_{r=-\infty}^{\infty} a_r x_{t-r}$$

This is convolution in statistical context and the coefficients a_t are called impulse response function and required absolute summability is given by

$$\sum_{t=-\infty}^{\infty} |a_t| < \infty$$

Then y_t exists as a limit in mean square and the infinite formula is

$$A(v) = \sum_{t=-\infty}^{\infty} a_t \exp\{-2\pi i\, vt\}$$

called frequency representation which is well defined.

The output spectrum is given by

$$f_y(v) = |A(v)|^2 f_x(v)$$

and autocovariance $\gamma_y(h) = \int_{-1/2}^{1/2} e^{2\pi i vh} |A(v)|^2 f_x(v)\,dv$

First difference and MA filters

$$\nabla x_t = x_t - x_{t-1} \text{ is an asymmetric filter.}$$

Symmetric filter $\qquad y_t = \dfrac{1}{2}(x_{t-6} + x_{t+6}) + \dfrac{1}{12}\sum_{r=-5}^{5} x_{t-r}$

Table 1.3   Characteristics of Symmetric MA and First Difference filters

	Geometry	*Frequency passed*	*Amplitude A(v)*	*Frequency response function*
Symmetric MA	Symmetric	High pass	Real: $1 - e^{-2\pi i v}$	Real
First difference	Asymmetric	Low pass	Real: $\dfrac{1}{12}\left[1 + \cos(12\pi v) + 2\sum_{k=1}^{5} \cos(2\pi vk)\right]$	Complex

Cross-spectra $f_{yx}(v) = A(v)f_x(v)$

$$A(v) = \frac{f_{yx}(v)}{f_x(v)} = \frac{C_{yx}(v)}{f_x(v)} - i\frac{q_{yx}(v)}{f_x(v)}$$

$$= |A(v)| \exp\{-i\phi_{yx}(v)\}$$

where amplitude filter is $|A(v)| = \sqrt{\dfrac{c_{yx}^2(v) + q_{yx}^2(v)}{f(v)}}$

and phase filter is $\qquad \phi_{yx}(v) = \tan^{-1}\left(-\dfrac{q_{yx}(v)}{c_{yx}(v)}\right)$

A simple delay filter is given by

$$y_t = Ax_{t-D}$$

A = amplifier $\quad D$ = delay time

$$f_{yx}(v) = Ae^{-2\pi i vD}f_x(v)$$

and amplitude is $|A|$ and phase is $\phi_{yx}(v) = -2\pi vD$ which is a linear function of frequency v.

Discrimination on the basis of S-wave spectra is very good with earthquake restricted < 0.7 Hz and explosions (nuclear) > 0.7 Hz. Also, power spectra are substantial above 1.5 Hz, whereas it is nearly zero for explosions. These figures suggest that ratio of P/S processes for earthquakes and explosions can give a good discrimination for these two categories.

Signal extraction

Suppose signal is contaminated with noise and we wish to extract the signal whose estimate is of the form

$$\hat{x}_t = \sum_{r=-\infty}^{\infty} a_r y_{t-r}$$

where the signal plus noise model is

$$y_t = x_t + z_t$$

$$\text{MSE} = E\left[\left(x_t - \sum_{r=-\infty}^{\infty} a_r y_{t-r}\right)^2\right]$$

Applying the orthogonality of signal and noise, we get

$$E\left[\left(x_t - \sum_{r=-\infty}^{\infty} a_r \, y_{t-r}\right) y_{t-s}\right] = 0$$

which leads to $\quad \sum_{r=-\infty}^{\infty} a_r \gamma_y(s-r) = \gamma_{xy}(s)$

Then, we have

$$A(v)f_y(v) = f_{xy}(v)$$

where $A(v)$ and optimal filter a_t are Fourier transform pairs.

We also have $\quad f_{xy}(v) = \overline{B(v)} f_x(v)$

$$f_y(v) = |B(v)|^2 f_x(v) + f_y(v)$$

So, $\qquad A(v) = \overline{B(v)} \Big/ \left(\left|\overline{B(v)}\right|^2 + \dfrac{f_v(v)}{f_x(v)}\right)$

where $f_x(v)/f_y(v)$ is the signal to noise ratio at frequency v.

If SNR is known $A(v)$ is computed; otherwise $a_t M = M^{-1}$

$\sum_{k=0}^{M-1} A(v_k) \times e^{2\pi i v_k t}$ is the estimated causal filter function.

Spectral analysis of multi-dimensional series

Multi-dimensional series of the form x_s, where $s = (s_1, s_2, ..., s_n)$ is an r-dimensional vector of spatial coordinates or a combination of space and time coordinates. Examples are a collection of temperature data over a rectangular field is a two-dimensional series, most of geological maps digitized over N-S, E-W or along any two orthogonal coordinates. The multi-dimensional wave number space is given by Fourier transform of autocovariance function.

$$f_x(v) = \sum_h \gamma_x(h) e^{-2\pi v' h}$$

The inverse result is

$$\gamma_x(h) = \int_{-1/2}^{1/2} f_x(v) e^{2\pi v' h} \, dv$$

Here, v is like frequency and cycle rate v_i per distance travelled s_i in the ith direction. Replacing the integral by summation, we have

$$f_x(v_1, v_2) = \sum_{h_1=-\infty}^{\infty} \sum_{h_2=-\infty}^{\infty} \gamma_x(h_1, h_2) e^{-2\pi i(v_1 h_1 + v_2 h_2)}$$

and $\quad \gamma_x(h_1, h_2) = \displaystyle\int_{-1/2}^{1/2} \int_{-1/2}^{1/2} f_x(v_1, v_2) e^{2\pi i(v_1 h_1 + v_2 h_2)} dv_1 \, dv_2$

in case $\quad r = 2.$

We define impulse response function s_1, s_2—the spatial filter output as

$$y_{s_1, s_2} = \sum_{u_1} \sum_{u_2} a_{u_1, u_2} x_{s_1 - u_1} x_{s_2 - u_2}$$

and spectral output is

$$f_y(v_1, v_2) = |A(v_1, v_2)|^2 f_x(v_1, v_2)$$

where $\quad A(v_1, v_2) = \displaystyle\sum_{u_1} \sum_{u_2} a_{u_1, u_2} e^{-2\pi i(v_1 u_1 + v_2 u_2)}$

which is analogous to the result obtained for one-dimensional case. Multi-dimensional DFT is a straightforward generalization of a one-dimensional case and sample wave numbers can be estimated over the grid with L_i's odd, as

$$\frac{2 L_1 L_2 \hat{f}_x(v_1, v_2)}{f_x(v_1, v_2)} \sim \chi^2_{2 L_1 L_2}$$

where $\quad \hat{f}_x(v_1, v_2) = (L_1 L_2)^{-1} \displaystyle\sum_{l_1, l_2} |X(v_1 + l_1/x_i, v_2 + l_2/n_2)|^2$

This result can give the confidence intervals or make approximate tests against a fixed assumed spectrum $f_0(v_1, v_2)$ (see Fig. 4.1).

1.6 SIMPLE TIME SERIES MODELS

Box and Jenkins (1970) have introduced backshift operator B, difference operator $\nabla = 1 - B$ and forward-shift operator F for easy representation and analysis of time series data. For example,

$$Bx_t = x_{t-1}, Fx_t = x_{t+1}$$

and $\quad \nabla x_t = x_t - x_{t-1} = (1 - B)x_t$

These quantities behave as any algebraic variable and can give correct results for all algebraic operations made in them. Here, we represent a few simple time series models and their B-J representations.

(i) *White noise ($z_t \sim WN(0, \sigma^2)$)*

If z_t is a white noise, then we have $E(z_t) = 0$, autocovariance

$$\gamma(h) = \begin{cases} \sigma^2; & h = 0 \\ 0; & h \neq 0 \end{cases}$$

and autocorrelation

$$\rho(h) = \begin{cases} 1; & h = 0 \\ 0; & h \neq 0 \end{cases} \text{ since } z_t, z_{t+h} \text{ are uncorrelated for } h \neq 0.$$

This has a flat spectral density $f(\lambda) = \sigma^2/2\pi; \quad -\pi < \lambda < \pi$.

It is called a white noise process since the spectral density is flat and each frequency contributes equally to the variance of the process.

(ii) *Autoregressive process of First Order: AR(1)*

Let $X_t = \phi X_{t-1} + z_t$ with $|\phi| < 1$, where $\{z_t\} \sim WN(0, \sigma^2)$.

Then, writing X_{t-1} as BX_t and shifting to the left-hand side, we get

$$(1 - \phi B)X_t = z_t$$

which is convertible to MA(∞) since $|\phi| < 1$.

It has mean $X_t = 0$ and variance $\gamma_x(0) = \dfrac{\sigma^2}{(1 - \phi^2)}$.

Note: variance $\gamma(0)$ is finite and X_t is stationary so long as $|\phi| < 1$ (stationarity condition). $\gamma_X(h) = \phi\gamma_X(h - 1) = \phi^{|h|}\gamma_X(0)$.

This gives $\rho(h) = \phi^{|h|}, h = 0, \pm 1, \ldots$

which is 1 for $h = 0$ and exponentially decreasing as $\log h$ increases, since $|\phi| < 0$. The spectral density is given by

$$f(\lambda) = \frac{1}{2\pi} \sum_{h=-\infty}^{\infty} e^{-ih\lambda} \gamma(h)$$

$$= \frac{\sigma^2}{2\pi(1 - \phi^2)} \left(1 + \sum_{h=1}^{\infty} \phi^h (e^{-ih\lambda} + e^{ih\lambda})\right)$$

$$= \frac{\sigma^2}{2\pi} (1 - 2\phi \cos \lambda + \phi^2)^{-1}$$

As expected for high values of ϕ, the spectral density is high at low frequencies and small for high frequencies. AR(1) can be inverted to give MA(∞) process since $|\phi| < 1$.

(iii)   MA(1) process

$$X_t = z_t + \theta z_{t-1} = (1 + \theta B)z_t, \; z_t \sim WN(0, \sigma^2)$$

with $|\phi| < 1$ (though this restriction is not necessary).

It has mean $EX_t = 0$ and finite variance $(1 + \theta^2)\sigma^2$, so that the process is always stationary. Its autocovariance function is

$$\gamma_X(t + h, t) = \begin{cases} \sigma^2(1 + \theta^2) & \text{if } h = 0 \\ \sigma^2\theta & \text{if } h = 0 \\ 0 & \text{if } h = 0 \end{cases}$$

and autocorrelations $\rho_x(h) = \begin{cases} 1 & \text{if } h = 0 \\ \theta/(1 + \theta^2) & \text{if } h = \pm 1 \\ 0 & \text{otherwise} \end{cases}$

The spectral density $f(\lambda) = \dfrac{\sigma^2}{2\pi}(1 + 2\theta \cos \lambda + \theta^2)$.

However, MA processes are not generally invertible and MA(1) is invertible to AR(∞) iff $|\rho_1| < 1/2$. Now, it is obvious that AR and MA processes are not competing models but are complementary. It would be prudent, therefore, to model a stationary and invertible time series $\{X_t\}$ as a mixed model ARMA (p, q) such that $(p + q)$ is minimum. This is the parsimony principle introduced by Box-Jenkins (1970, 1976).

(iv)   ARMA (1,1) process

If $\{X_t\}$ is a causal and invertible ARMA(1,1) process

$$X_t - \phi X_{t-1} = z_t + \theta \hat{z}_t \text{ with } |\phi| < 1 \text{ and } |\theta| < 1$$

Then its mean is zero and variance $\gamma_0 = \phi\gamma_1 + \sigma^2(1 + \theta + \phi)$

and $\gamma_1 = \phi\gamma_0 + \theta\sigma^2$

This yields $\gamma_0 = (1 + 2\phi\theta + \theta^2)\, \sigma^2/(1 - \phi^2)$

and $\gamma_1 = (\phi + \theta)\,(1 + \theta\phi)\,\sigma_a^2/(1 - \phi^2)$

and all higher autocovariances are zero.

The autocorrelation is

$$\rho_1 = (\phi + \theta)(1 + \phi\theta)/(1 + 2\phi\theta + \theta^2)$$

where $(\theta + \phi) \neq 0$.

Its spectral density is given by

$$f_x(\lambda) = \frac{\sigma^2 \left|\theta\left(e^{-i\lambda}\right)\right|^2}{2\pi \left|f\left(e^{-i\lambda}\right)\right|^2}, -\pi \leq \pi$$

which is the ratio of trigonometric polynomials and hence called a rational spectral density.

For ARMA (1,1) process, we can simplify spectral density $f_x(\lambda)$ to

$$f_x(\lambda) = \frac{\sigma^2 (1 + \theta^2 + 2\theta \cos \lambda)}{2\pi(1 + \phi^2 + 2\phi \cos \lambda)}$$

Box-Jenkins (1970) introduced multiplicative seasonal models where the seasonal part is multiplied to the non-seasonal part. Therefore logarithmic transformation of variable, i.e. $X_t = \log X_t^*$ linearly separates the non-seasonal part from the seasonal and each part can be modelled separately as well as independently to yield the final model. Further, the difference operator $\nabla = (1 - B)$ can be utilized to convert a non-stationary series to a stationary series which can be modelled as an ARMA (p, q) process $\phi_p(B)X_t = \theta_q(B)z_t$, where $\phi_p(B)$ and $\theta_q(B)$s are degree p and q polynomials in back-shift operator B.

The original non-stationary series has the model ARIMA (p, d, q) where $d > 0$ is the degree of differencing necessary to obtain stationarity condition (i.e. minimum variance occurs for the d-differenced series). In the above models, X_t is the time series and z_t is a white noise process. If $\{X_t\}$ is stationary, then $\phi(B)$ has an inverse $\phi^{-1}(B)$ and hence we can convert an ARMA (p, q) process to an equivalent purely MA(∞) process as

$$X_t = \phi^{-1}(B)\,\theta(B)z_t$$

$$= \psi(B)z_t$$

where $\psi(B)$ is the transfer function for converting white noise process z_t through a linear filter to the time series X_t. This method is useful in forecasting and updating the forecasts. Similarly, if the MA polynomial $\theta_q(B)$ is invertible (i.e. $\theta_q^{-1}(B)$ exists), then we can invert an ARMA (p, q) process to an equivalent purely AR (∞) process as

$$\pi(B)X_t = \theta_q^{-1}(B)\phi_p(B)X_t = z_t$$

where $\pi(B) = \theta_q^{-1}(B)\phi_p(B)$.

This method is useful in estimating the autoregressive coefficients that are statistically significant and finding the order of the equivalent AR process. The principle of parsimony (Box-Jenkins concept) then suggest that stationary time series $\{X_t\}$ be modelled as ARMA (p, q) model with the constraint that the number of parameters in model $(p + q)$ will be minimum. Although it is not absolutely necessary, but we can restrict $(p + q) \leq 2$ to obtain simple stationary time series models which are highly flexible as to the values of parameters, $-1 < \phi < 1$ and $-1 < \theta < 1$, as well as the different types of candidate models.

Thus, we can have a wide range AR (1), AR (2), MA (1), MA (2) and ARMA (1,1) models with only two parameters ϕ_i and/or θ_j, with i, j either 1 or 2 such that $i + j \leq 2$.

If we add the degree of differencing $0 \leq d \leq 2$, then the number of candidate non-stationary/stationary type of time series models rises to fifteen. Therefore, the Box-Jenkins modelling techniques are simplest, flexible and cover a wide range of possible time series obtained through natural observations and experiments. In addition, we can use non-linear transformation of data to model the time series using the linear ARMA (p, q) or ARIMA (p, d, q) process.

1.7 MODEL VALIDATION AND INFERENCES

Although the autocorrelation (a.c.f.) and spectral density (f) are very useful tools for identifying the time series model, better criteria are available for model identification. It is well known for multivariate statistics that multiple (auto) correlation coefficient does not provide the true correlation between output X_t and past values X_{t-1}, X_{t-2}, etc. What we require is partial autocorrelation function between X_t and X_{t-k} for each $0 < k < N/4$.

If $\{X_t\}$ is stationary with $t = 1$ to N, we compute the mean function

$\hat{\mu} = \bar{x} = \dfrac{1}{N}\sum_{i=1}^{N}x_i$ and autocovariance at lag k is given by

$$\hat{r}_k = c_k = \frac{1}{N}\sum_{k+1}^{N}(x_i - \bar{x})(x_{i-k} - \bar{x}), \, k < (N/4)$$

The autocorrelation $\hat{r}_k = c_k/c_0$.

If this process is stationary and Gaussian, then $\rho_k = 0$ for all $k >$ some large K.

$$Cov[r_k, r_{k-s}] = \frac{1}{N} \sum_{K+s}^{K} \rho_i \rho_{i-s}$$

Putting $s = 0$, $Var[r_k] = \frac{1}{n} \sum_{-K}^{K} \rho_i^2$ and for large N, $\rho_k = 0$ and r_k will be approximately normally distributed.

Thus, $$Var[r_k] = \frac{1}{N}\left(1 + 2\sum_1^K r_i^2\right)$$

and square-root of this is the large-lag standard error of r_k. Approximate 95% confidence interval is given by

$$\pm 2SE[r_k] = \pm 2\frac{1}{N}(1+2)\sum_1^K r_i^2)^{1/2}$$

The autocorrelation can be arranged in matrix form as

$$\rho_N = \begin{bmatrix} 1 & \rho_1 & \cdots & \rho_{N-1} \\ \rho_1 & 1 & \rho_1 & \rho_{N-2} \\ \cdots & \cdots & \cdots & \cdots \\ \rho_{N-1} & \rho_{N-2} & \rho_{N-3} & 1 \end{bmatrix}$$

which is positive definite.

Thus, acf and acvf suffer many constraints. For example, if

$$\begin{vmatrix} 1 & \rho_1 & \rho_2 \\ \rho_1 & 1 & \rho_1 \\ \rho_2 & \rho_1 & 1 \end{vmatrix} > 0$$

implies $(1 - \rho_2)(1 + \rho_2 - 2\rho_1^2) > 0$.
Since $1 > \rho_2$, the constraint becomes $\rho_2 > 2\rho_1^2 - 1$.

The partial autocorrelation function $\{\phi_{kk}\}$ ($k = 1, 2 \ldots$) is given by

$$\phi_{kk} = |p^*_k| / |p|$$

where p_k is a.c. matrix and p^*_k is the matrix with last column replaced by the column ρ such that $\rho^T = [\rho_1, \rho_2, \ldots \rho_k]$.

So we obtain $\phi_{11} = \rho_1$ and $\phi_{22} = \begin{vmatrix} 1 & \rho_1 \\ \rho_1 & \rho_2 \end{vmatrix} / \begin{vmatrix} 1 & \rho_1 \\ \rho_1 & 1 \end{vmatrix} = \dfrac{\rho_2 - \rho_1^2}{1 - \rho_1^2}$ and so on.

The estimate of ϕ_{kk} is $\hat{\phi}_{kk}$, where ρ_i is replaced by sample auto-correlation coefficient r_i.

According to Quenouille $Var\left[\hat{\phi}_{kk}\right] = \dfrac{1}{N}$ and $SE\left[\hat{\phi}_{kk}\right] = \dfrac{1}{\sqrt{N}}$ and 95% confidence belt is given by $\pm 2/\sqrt{N}$.

If $\hat{\phi}_{pp}(p < k)$ lies beyond $\pm 2SE\left[\hat{\phi}_{kk}\right]$, then these are statistically significant, which means X_t is correlated significantly with $X_{t-1}, ..., X_{t-p}$ and not correlated with $X_{t-(p+1)}, ..., X_{t-k}$. Then the model is AR($p$) process.

Similarly, if a.c. $\hat{r}_q(q < k)$ are significant at 95% confidence level, then the model is MA (q) process. If some of the r_k and ϕ_{kk} are significant at 95% confidence level, then the process is $WN(0, \sigma^2)$. If r_1 and ϕ_{11} are significant at 95% level and these exponentially decrease (not cut off) for larger lags, then the model is ARMA(1,1).

Once the model is identified, the signal (trend) can then be estimated and removed. If the model is adequate then the residual series should be a white noise with mean zero and variance σ^2(error variance), which is useful in finding the standard error of the forecasts. In identification of model, we wish to use the parsimony principle and penalize models having larger number of parameters. These penalty criteria include.
   (i)   FPE (finite prediction error)
   (ii) AICC (bias corrected AIC)

(i) FPE

If $\{X_t\}$ and $\{Y_t\}$ are two independent realizations of an AR(p) process, we can estimate parameters $\hat{\phi}_i, i = 1, ..., p$ for $\{X_t\}$ data and compute one-step predictor Y_{N+1} as

$$\hat{\phi}_i Y_n + \cdots + \hat{\phi}_p Y_{N+1-p}.$$

Then, the mean square prediction error $= \sigma^2\left(1 + \dfrac{p}{N}\right).$

The FPE is given by $FPE_p = \hat{\sigma}^2\, \dfrac{N+p}{N-p}$

where σ^2 is replaced by the maximum likelihood estimate $N\hat{\sigma}^2/(N-p)$. We choose a p value, which is minimum over p.

(ii) AIC or AICC

It is more generally applicable than FPE and is based on the Kulbak-Leibler discrepancy between the true and approximate. We minimize

$$\text{AIC} = -2 \ln L(\phi_p, \theta_q, S(\phi_p, \theta_q)/N) + 2(p + q + 1).$$

$$\text{AICC} = -2 \ln L(\phi_p, \theta_q, S(\phi_p, \theta_q)/N) + 2(p + q + 1)N/(N - p - q - 2).$$

using the estimated values of parameter ϕ_p and θ_q, etc., and then compute error variance σ^2 for the accepted model.

(iii) BIC

BIC is similar to AIC or AICC and is given by

$$\text{BIC} = (n - p - q) \ln\left[n\hat{\sigma}^2/(N - p - q)\right] + N(1 + \ln\sqrt{2\pi})$$

$$+ (p + q) \ln\left[\left(\sum\nolimits_{t=1}^{N} X_t^2 - N\hat{\sigma}^2\right)\Big/(p + q)\right]$$

Max $\hat{\sigma}^2$ is MLE of white noise variance. BIC is a consistent estimator, i.e. $\hat{p} \to p$ and $\hat{q} \to q$ with probably 1 as $N \to \infty$, which is not true for AIC or AICC.

1.8 STATE-SPACE MODEL AND KALMAN FILTERING

State-space Models

State-space models are of current interest of concern for navigation and control of rockets and Kalman filter is a general method for obtaining the optimal recursive estimates of current state of a dynamic system and its variances. It is thought that any measurement includes signal plus contaminated noise. The state-space is that signal comprises a linear combination of a set of variables (state variables) continuing the state vector at time t. The state variables include current time underlying level and current seasonal factor (if any). We denote $(m \times 1)$ state vector by θ_t and write observation

$$X_t = h_t \cdot \theta_t + n_t \qquad \text{(observation/measurement equation)}$$

where $(m \times 1)$ vector h_t is known and n_t is observation error. However, θ_t cannot be observed directly but we assume θ_t can be updated by equation

$$\theta_t = G_t \theta_{t-1} + w_t \text{ (system/transition equation)}$$

where $(m \times m)$ matrix G_t is known, and w_t denotes a vector of deviations. The errors n_t and w_t are uncorrelated with each other for all t and also both are serially uncorrelated. We assume n_t is $N(0, \sigma_n^2)$ and w_t to be multivariate normal (mnd) with mean zero vector and covariance matrix w_t. We can easily express this state-space equation where X_t is a vector and h_t is a matrix and n_t of appropriate length. We can also add the linear combination of explanatory (or exogenous) variables to the RHS of equation. If seasonal effects are multiplied, it would be best to use logarithmic transformation to make the seasonal effects additive. This allows local trend as seasonality to be updated rather than time using the system equation.

Steady Model

We can write the observation as $X_t = \mu + n_t$ and the current level updates as

$$\mu_t = \mu_{t-1} + w$$

which is a random walk.

The state vector θ_t is a single state variable μ_t (scalar), while h_t and G_t are also scalar constants. The errors are assumed normal with variances σ_n^2 and σ_w^2 and ratio σ_w^2 / σ_n^2 is called signal-to-noise ratio. If $\sigma_w^2 = 0$, then $\mu_t = \mu_{t-1}$, which is a constant mean model (trivial) in contrast to linear growth model for varying μ_t. The simple exponential smoothing produces optimal forecasts for an ARIMA $(0,1,1)$ model as also for this steady model. Therefore, a steady model has the following characteristics:

(i) no long-term trend; and
(ii) no seasonality but some short-term correlations.

Linear Growth Model

We have three equations

$$X_t = \mu + n_t \qquad \text{(observation)}$$

$$\mu_t = \mu_{t-1} + B_{t-1} + w_{1,t} \qquad \text{(transition equation)}$$

$$B_t = B_t + w_{2,t} \qquad \text{(transition equation)}$$

The vector $\theta'_t = (\mu_t, B_t)$ has two components, which are 'local level μ_t'
and 'local trend B_t'. We then have $h'_t = (1,0)$ $G_t = \begin{bmatrix} 1 & 1 \\ 0 & 1 \end{bmatrix}$.

If $w_{1,t}$ and $w_{2,t}$ have both variances to be zeros, then the trend is constant (deterministic) and called a global linear trend model, which is rather unlikely in case of real phenomena. We generally prefer a *local* linear trend, where the trend may change. It is easier to obtain a variety of trend models for this general state-space model rather than the ARIMA class of models. Second-differences of X_t are stationary and have some a.c.f. as an MA (2) model. Therefore, a two-parameter (updating level and also trend) exponential smoothing would be optimal for an ARIMA (0,2,2) model as well as for the linear growth model. Seasonality(s) can be introduced as follows to give structural models:

$$X_t = \mu_t + i_t + n_t; \qquad \mu_t = \mu_{t-1} + B_{t-1} + w_{1,t}$$

$$B_t = B_{t-1} + w_{2,t}; \qquad i_t = -\sum_{j=1}^{(s-1)} i_{t-j} + w_{3,t}$$

with state-vector having $(s + 2)$ components. Further extension for interventional and explanatory variables can be made in these structural models.

Kalman Filter

Kalman filter is a general method to estimate the state vector θ_t by obtaining a set of equations which allows us to recursively compute the estimate of θ_t when a new observation becomes available. The two stages for updating procedure are called:

   (i) prediction stage; and
   (ii) updating stage.

Prediction stage has measurements up to time $t-1$ and then $\hat{\theta}_{t-1}$ is the best estimator for θ_{t-1} (i.e. minimum MSE estimator). Let the covariance matrix of $\hat{\theta}_{t-1}$ be denoted at P_{t-1} is also computed. We can forecast θ_t from information at $(t-1)$ as $\hat{\theta}_{t/t-1}$.

However, w_t is unknown at time $(t-1)$ and

$$\hat{\theta}_{t|t-1} = G_t \hat{\theta}_{t-1} \qquad \qquad \text{(Prediction equations)}$$

and its covariance matrix is $P_{t|t-1} = G_t P_{t-1} G'_t + W_t$.

Updating stage: When a new observation X_t is available at time t, estimator θ_t can be modified using this extra information.

The predictor error is $e_t = X_t - h'_t \hat{\theta}_{t|t-1}$

And updating equations are given by

$$\theta'_t = \hat{\theta}_{t|t-1} + K_t e_t$$

$$P_t = P_{t|t-1} - K_t h'_t P_{t|t-1} \qquad \qquad \text{(Updating equations)}$$

where $K_t = P_{t|t-1} h_t / \left[h'_t P_{t|t-1} h_t + s_n^2 \right]$ is the Kalman gain matrix (in univariate case, it is a vector). The main advantage is that these equations are recursive so that a large memory is not required for updating. Another advantage is that such an equation converges quickly ($\equiv$ exponential smoothing operation) if there is a constant structure model or it follows systemic movements through time in case of seasonal structure models. Even multiplicative seasonal models (locally non-linear) can be solved by extending the Kalman filter to locally linear approximation to the model.

The linear growth model can also be solved by using Kalman filter procedure. Suppose, from data up to time $(t-1)$ we obtain estimates $\hat{\mu}_{t-1}$ and $\hat{\beta}_{t-1}$ for level and trend.

At time $(t-1)$, the best forecasts of $w_{1,t}$ and $w_{2,t}$ are both zero, so that the best forecasts of μ_t and β_t are given by

$$\hat{\mu}_{t|t-1} = \hat{\mu}_{t-1} + \hat{\beta}_{t-1}$$

and $\qquad \hat{\beta}_{t|t-1} = \hat{\beta}_{t-1}$

When X_t becomes available, we find $e_t = X_t - \hat{\mu}_{t|t-1}$ and then

$$\hat{\mu}_t = \hat{\mu}_{t|t-1} + c_{1,t}\, e_i \qquad\qquad \hat{\beta} = \hat{\beta}_{t|t-1} + c_{2,t}\, e_t$$
$$= \hat{\mu}_{t-1} + \hat{\beta}_{t-1} + c_{1,t}\, e_t \qquad \text{and} \qquad = \hat{\beta}_{t-1} + c_{2,t}\, e_t$$

where $c_{1,t}\, e_t$ and $c_{2,t} e_t$ are the elements of Kalman gain matrix (here a 2×1 vector). K_t then can be evaluated after some algebraic computations. These are similar to 2-parameter Holt-Winters updating where L_t, T_t are denoted as level and trend. We had (H-W equations)

$$L_t = \alpha X_t + (1 - \alpha)\,(L_{t-1} + T_{t-1})$$
$$= L_{t-1} + T_{t-1} + \alpha e_t$$

where $e_t = X_t - [L_{t-1} + T_{t-1}]$.

In the steady state as $t \to \infty$, $c \to$ constant corresponding to smoothing parameter α which shows that 2-parameter H-W is optimal for linear growth model. We can initialize the two state variables for the first two observations as

$$\hat{\mu}_2 = X_2 \quad \text{and} \quad \hat{\beta}_2 = X_2 - X_1$$

1.9 PROCEDURES FOR TIME SERIES DATA ANALYSIS

Time series data analysis is not easy as several complex questions need to be answered before deciding on the appropriate candidate models for detailed statistical analysis. A systematic structure procedure would be useful, so that steps required are performed only once within a loop run and repeated many times during the analyses. The analysis is done only for homogeneous time series with two discontinuities (known/inferred).

(i) The first step is whether the series is non-stationary or stationary and in case of the former, how to convert a non-stationary series to a stationary one (with homogeneous second-order properties). It is suggested that the large data set be divided in three equal parts and means, variances, and auto-covariances computed for each part separately to give $\bar{x}_i$, s_i^2, $\gamma_i(h)$ for $i = 1, 2, 3$. If there is no difference in s_i^2, then homogeneity in variances is acceptable and we test homogeneity of means $\bar{x}_i$. If homogeneity in the means is acceptable, then we can use the standard procedure of modelling such as seasonality, ARMA (p, q), etc. However, if homogeneity of means $\bar{x}_i$ is not acceptable, then there can be linear, quadratic or trigonometric time trends and/or discontinuities. These non-stationarities can be tackled using polynomial trend removal, differencing/fractal differencing or intervention models, respectively. On the other hand, if variances s_i^2 are increasing with values of $i (= 1, 2, 3)$, then we require transformation of the variable either by logarithmic ($x_t = \ln y_t$) or by power transformation of Box-

$$\text{Cox type } \left(x_t = \begin{cases} (y_t^\lambda - 1) / \lambda, & \lambda \neq 0 \\ \log y_t & \lambda \neq 0 \end{cases} \right).$$

These non-linear transformations induce homogeneity in variances, covariances as well as improve the normality and linearity of the random variables.

(ii) The next step is to find whether the series is non-seasonal, multiplicative seasonal or additive seasonal. By plotting the homogeneous and stationary time series data against time, it is easy to recognize these three cases. Multiplicative seasonality can be handled by logarithmic transformation of the data and modelling the transformed data separately and independently (by suitable seasonal differencing) for modelling non-seasonal and seasonal components. These two independent models are finally recombined to set forecast values, etc. If the series is additive seasonal, then a state-space model with Kalman filters and smoothers may be used to remove the seasonalities and the time trend is non-seasonal for modelling purposes. An alternative approach is to recognize the seasonalities by Fourier (frequency domain) analysis and remove them for modelling the non-seasonal part. Frequency domain approach is specially meaningful in case of the study of earthquake waves, brain waves, wind vibrations in wings of aeroplane, temperature variations during day/month, and in environmental, social economic sciences.

(iii) The observed time series may be the resultant of superposition of two or more geological processes such as tectonics, minerali-zations, diagenesis, etc. The resulting superposed models would be very complex and it would be of interest to decompose them into their simple ARMA (p, q) or ARIMA (p, d, q) components. If so, these simpler models help in the genesis of the observed geological phenomena with respect to their time/spatial history, as well. The analyses and inferences would be useful for the possible control of the natural geological processes and also for exploration of valuable mineral resources, etc.

1.10 EXAMPLES

Example 1

Closing stock prices for 50 days were noted for a given stock and the aim was to find a possible time series model. We obtain $\bar{x} = 288, c_0 = 4, c_1 = -1$, so $r_1 = -0.25$, which does not seem to be significant (nor ϕ_{11} is significant). Therefore, a white noise model seems acceptable and we cannot forecast the next day's closing price at the stock exchange.

Example 2

Simulation of AR (1) model with $\phi = 0.9$ gave the following acf values based on $N = 50$ and only ϕ_{11} was found to be significant and $\phi_{kk} = 0$ for $k > 1$.

Table 1.4

k	1	2	3	4	5	6	7	8	9	10
r_k	0.84	0.73	0.61	0.54	0.47	0.46	0.38	0.29	0.17	0.05
f_{kk}	0.84	0.08								

Therefore, based on significance of ϕ_{kk} and exponential decrease of r_k, we conclude that the model is AR(1) (which was, in fact, used to generate these 50 data).

Example 3

Simulation of MA (1) model with $\theta = -0.6$ gave fitting acf $r_1 = -.507$ and the higher lag autocorrelations were very small. Variance $[r_k] = \dfrac{1}{N}(1 + 2r_1^2) = 0.03$ with 95% confidence limits $\pm 2 \times 0.17 = \pm 0.34$. Thus, r_1 was significant with cutoff from r_2 onwards. So, we can accept a model of MA (1) as reasonable for these data.

Example 4

Recently, global warming has been of great concern and the yearly average temperatures have risen for 1900 to 1997, although it was levelled off around 1935-1980. The question may be asked whether there is uniform trend in the warming process and is it caused by human activity. If so, how we can control it? For example, CFC input or reducing coal/oil for energy.

Example 5

Melting glaciers deposit coarser sediments (sand and silt) in summer and finer clays in winter, giving couplets for 1 year (varves). Such varves have been recorded in Massachusetts for 634 years, beginning with 11,834 years ago. The thickness distribution of varves is approximately lognormal. Hence, log-transforms have homogeneous variances and a log-transformed thickness versus years could be more useful than thickness against year for the data.

Example 6 (Smoothing)

Time series data x_t are often choppy due to presence of large random noise, i.e.

$$x_t = f(t) + z_t.$$

The function $f(t)$ may be considered a smooth function, which could be:

(i) polynomial of order p

(ii) periodic trigonometric regressional with cos and sine functions

(iii) kernel smoothing $f(t) = \sum_{i=1}^{n} z_t(i)$, where $z_t(i) = K\dfrac{t-i}{b}$

$$\Big/ \sum_{i=1}^{n} \frac{k(t-j)}{b}. \text{ Typically, } K(z_n) \sim N(0,1)$$

(iv) by (piecewise) cubic splines $\sum_{t=1}^{n}[x_t - f(t)]^2 + \lambda \int (f''(t))^2 dt$,

where $\lambda > 0$ controls the degree of smoothness.

Example 7   *Stationarity by polynomial regression versus differencing*

We can achieve stationarity in a time series by finding the trend and removing it from the data. This is achieved by either using a differencing operator to obtain the minimum variance of differenced series or by subtracting the accepted polynomial trend function from the original data.

The main advantages of use of differencing operator is that:

(i) It does not assume that a trend is a continuous function of t and, hence, useful where data have piecewise continuous trend also.

(ii) In differencing, we do not have the problem of testing the significance of the order of the polynomial using F-test, which is not applicable as the time series data are usually correlated.

Example 8   *Earthquakes and mining explosions*

We record arrivals of P and S waves against time. The amplitude versus time record shows sharp oscillations and the amplitude ratio of P and S waves for earthquakes is about 0.5, whereas for mining explosions, it is about 1.0. These two phenomena can, therefore, be easily discriminated. The question is: What about discrimination of earthquakes and nuclear explosions, which are of similar energy levels?

Example 9

Assay value distributions sampled at uniform distances along a drill core from an ore deposit can be modelled as one-dimensional spatial/time series X_t, where t represents the length along the drill (usually vertical) core.

(i) Is the data continuous or does it have discontinuities due to the presence of faults/(gangue) host rocks?

(ii) Does the data show stationarity in assay value distributions? If not, how to test it?

(iii) Is the variance of assay values homogeneous over the whole drill core? If not, can be suggested as a transformation to make variances homogeneous and variates approximately Gaussian?

Hint: Divide the drill core assay data into upper, middle and lower parts and determine the mean of each part separately. If $\bar{x}_1 = \bar{x}_2 = \bar{x}_3$, then there are no discontinuities and the stationarity may be assumed. However, if $\bar{x}_1 \neq \bar{x}_2 = \bar{x}_3$, then there could be fault or gangue zones. If $\bar{x}_1 > \bar{x}_2 > \bar{x}_3$ or $\bar{x}_1 < \bar{x}_2 < \bar{x}_3$, then there may be linear trend, or else we may think of quadratic trend in assays. Compute sample variances for each zone s_1^2, s_2^2, s_3^2 and if these are equal, no transformation of assay data is required. However, if s_1^2 is proportional to $\bar{x}_i$, then log x transformation could homogenize the variances as well as make assay data approximately Gaussian. Outliers, if any, should be recognized and discarded, before making data analysis.

Example 10

Usually, drill core logs are used for stratigraphic correlations over the drill holes. Can we obtain good correlations by cross-correlation of some property over two drill holes?

Ans: Each drill core may have several discontinuities due to unconformities, paraconformities, faultings, etc., and a continuous sequence is not guaranteed. In addition, sedimentation rates and supplies at two drill hole sites could be different and also variable at different parts of the core length (geological time). Hence, cross-correlation by time series would be error prone.

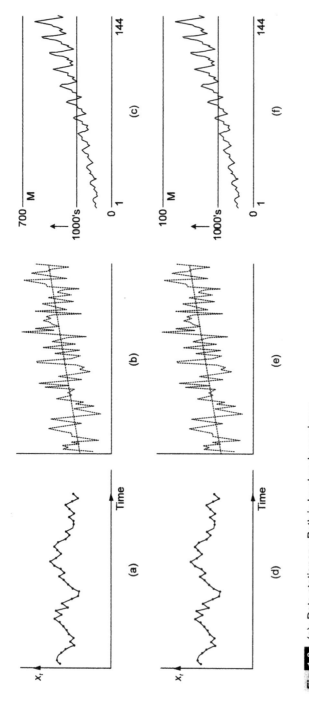

Fig. 1.3 (a): Data stationary: Both in level and covariance

(b): Data non-stationary in level and covariance is stationary (Non-stationary trend may be removed by polynomial trend subtraction or by differencing)

(c): Data non-stationary in level as well as in covariances (which can be made stationary by logarithmic transformation and trend removal by differencing)

(d): Intervention at time t_0 due to faulting/policy change

(e): Non-linear time series with sudden bursts/volatility (such as in cases of earthquakes, nuclear explosions, stock market prices, etc.)

(f): Nuclear explosions, stock market prices

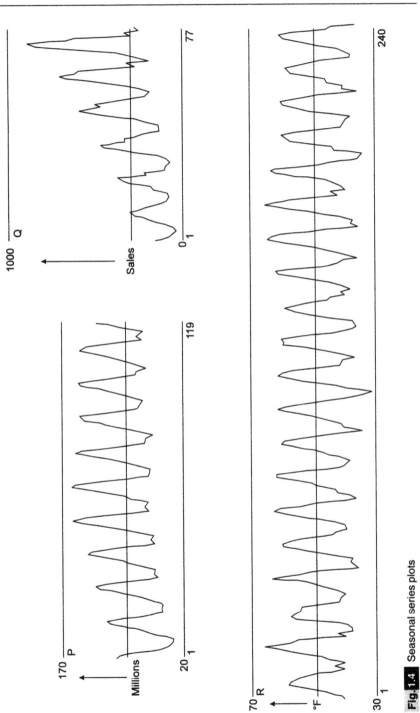

Fig. 1.4 Seasonal series plots

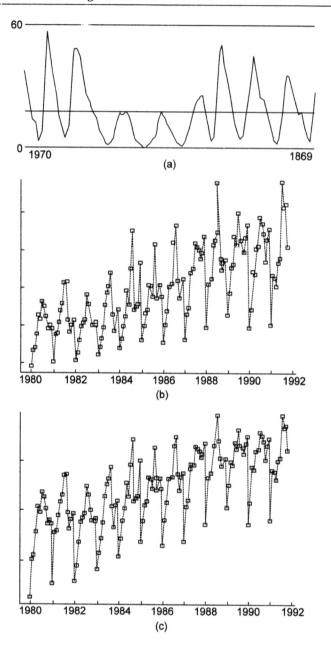

Fig. 1.5 (a): Stationary seasonal data with constant covariance (the annual sunspot numbers)

(b): Non-stationary seasonal data with increasing covariance

(c): Data after log-transform of (b) so that covariance becomes constant

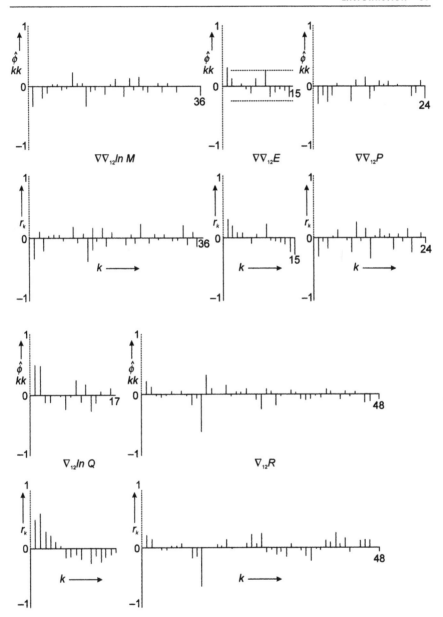

Fig. 1.6 Estimated functions for differenced seasonal series

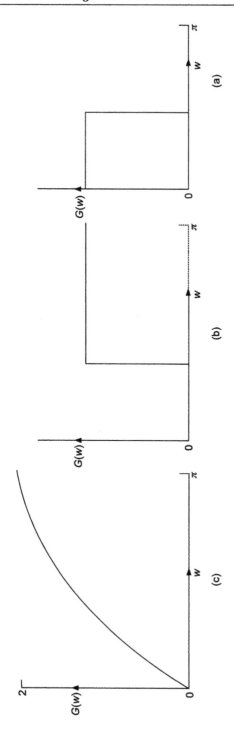

 Fig. 1.7   Ideal filters
(a) Low-pass filter or trend estimator
(b) High-pass filter or trend eliminator
(c) Gain diagram for the difference operator

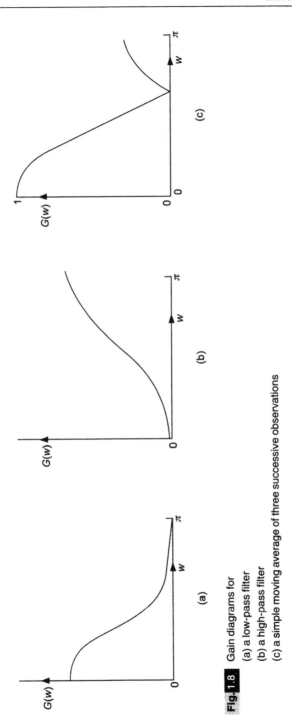

Fig. 1.8 Gain diagrams for
(a) a low-pass filter
(b) a high-pass filter
(c) a simple moving average of three successive observations

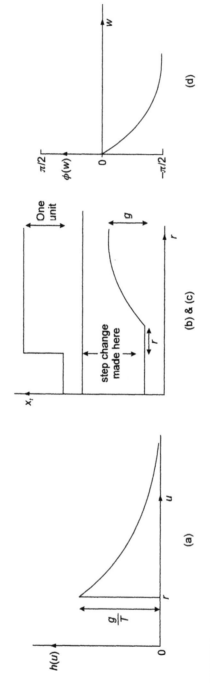

Fig. 1.9 (a) Delayed exponential response to a unit step change in input, showing graphs of impulse response function
(b) & (c) input and output, respectively
(d) Phase diagram of a simple exponential response function

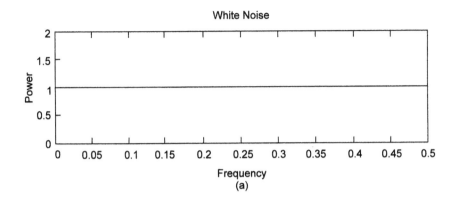

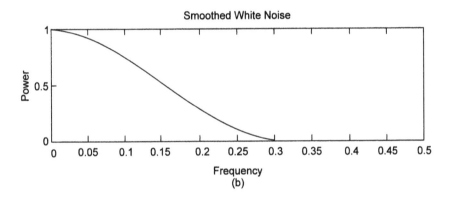

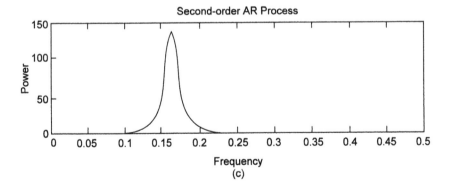

Fig. 1.10 (a) Theoretical spectra of white noise
(b) Smooth white noise
(c) Second-order AR process

1.11 EXERCISES

1. If X and Y are two random variables and $E(Y) = \mu$ and $\sigma_\gamma^2 < \infty$
   (a) show that $E(Y - \mu)^2$ is minimum.
   (b) If random variables $f(x)$ minimizes $E[(Y - f(X))]^2$, then $f(X) = E[Y \mid X]$
   (c) The above minimizing function $f(X)$ is also $f(X) = E(Y \mid X)$.

2. Let $\{z_t\}$ be a sequence of independent normal random variables, each with mean 0 and variance σ^2 and a, b, c are real constants. Then, check which of these are stationary in their mean and ACVF:
   (a) $X_t = z_1 \cos(ct) + z_2 \sin(ct)$
   (b) $X_t = z_t \cos(ct) + z_{t-1} \sin(ct)$
   (c) $X_t = a + bz_t + cz_{t-1}$
   (d) $X_t = z_t z_{t-1}$

3. Let $m_t = \sum_{k=0}^{p} c_k t^k, t = 0, \pm 1, \dots$.

   Then, ∇m_t is a polynomial of degree $p - 1$ in t and therefore $\nabla^{p+1} m_t = 0$.

4. Show a linear filter $\{a_j\}$ passes an arbitrary polynomial of degree k that is

$$mt = \sum_j a_j m_{t-j}$$

   for $m_t = c_0 + c_1 t + \dots + c_k t^k$ iff $\sum_j a_j = 1$ and $\sum_j j^r a_j = 0$ for $r = 1$ to k.

   Show that for passage of third degree polynomial trends with distortion, a 15-point MA filter $\{a_j\}$ is required.

5. Let $\{Y_t\}$ be a stationary process with mean zero and a, b are constants.

   If $\{X_t\} = a + bt + s_t + Y_t$, s_t has period 12, show that $\nabla\nabla_{12} X_t = (1 - B)(1 - B^{12}) X_t$ is stationary. Find its autocorrelation function.

   If $\{X_t\} = (a + bt)s_t + Y_t$, show that $\nabla_{12}^2 X_t = (1 - B^{12})^2 X_t$ is stationary. Find its autocorrelation function.

6. Show that $\underset{=N}{P}$ is positive definite, using linear combination $\zeta_1 = $

$$\sum_{i=1}^{N} v_i x_i \text{ of constants } v_i, i = 1, 2, \dots, N, \text{ not all zero.}$$

Hint: Find $Var(\zeta_1) = v'P_N v\sigma_z^2$, which is always positive since $\sigma_z^2 > 0$.

7. Find the mean, autocovariance, autocorrelation for the simple moving average process

$$x_t = \frac{1}{3}(z_{t-1} - z_t + z_{t+1})$$

where $z_t \sim N(0,1)$.

Ans.

$$E(X_t) = 0, \ \gamma(s, t) = \begin{cases} 3/9 & s = t \\ 2/9 & |s - t| = 1 \\ 1/9 & |s - 2| = 2 \\ 0 & |s - t| \geq 3 \end{cases}, \ \rho(s, t) = \begin{cases} 1 & s = t \\ 2/3 & |s - t| = 1 \\ 1/3 & |s - t| = 2 \\ 0 & |s - t| \geq 3 \end{cases}$$

8. Suppose the time series X_t has a signal $\cos(2\pi t/50)$ but is contaminated by white noise $z_t \sim N(0, 25)$. Find the mean value.

Ans. $E(X_t) = \cos(2\pi t/50)$

9. Find the mean autocovariance and autocorrelation for the AR(1) process $X_t = .8X_{t-1} + z_t, \ z_t \sim WN(0,1)$.

$$E(X_t) = 0, \ \gamma(s, t) = \begin{cases} 1/0.36 & s = t \\ .8/.36 & |s - t| = 1 \\ 0 & |s - t| > 2 \end{cases}, \ \rho(s, t) = \begin{cases} 1 & s = t \\ .8 & |s - t| = 1 \\ 0 & |s - t| > 1 \end{cases}$$

10. If X_t is weakly stationary with mean μ and autocovariance $\gamma(h)$, $h = 0, \pm 1, \dots$

(a) Show that its sample mean $\bar{x} = \dfrac{1}{N}\displaystyle\sum_{i=1}^{N} x_t$ is unbiased.

(b) Show variance of sample mean $E[(\bar{x} - \mu)^2]$ is the sum of

$$\frac{1}{N}\gamma(0) + \frac{1}{2N}\sum_{h=1}^{n}\left(1 - \frac{h}{N}\right)\gamma(h).$$

(c) If x_t are independent, then $E[(\bar{x} - \mu)^2] = \dfrac{\sigma^2}{N}$ as required.

11. A real-valued function $f(t)$, defined on integers, is non-negative definite if

$$\sum_{s=1}^{N} \sum_{t=1}^{N} a_s f(s-t) a_t \geq 0.$$

For matrix $F\{f(s-t), s, t = 1, 2, \dots N\}$, this implies $a'\,Fa \geq 0$ for all vector a.

(i) Prove $\gamma(h)$ of stationary process is a non-negative definite function.

(ii) Verify that sample autocovariance function $\gamma(h)$ is also a non-negative definite function.

References

1. Akaike, H., 1967, Some problems in the application of the cross-spectral method. In: *Spectral Analysis of Time Series*, (ed. B. Harris), Wiley, New York.

2. Anderson, O.D., 1976, *Time Series Analysis and Forecasting*, Butterworth, p. 182.

3. Anderson, T.W., 1971, *Statistical Analysis of Time Series*, Wiley, New York.

4. Bloomfield, P., 1976, *Fourier Analysis of Time Series*, Wiley, New York.

5. Box, G.E.P. and Cox, D.R., 1964, An Analysis of Transformations, *J.R. Statistical Soc.*, B26: pp. 211-252.

6. Box, G.E.P. and Jenkins, G.M., 1970, 1976, *Time Series Analysis: Forecasting and Control*, Holden and Day, San Francisco, p. 575.

7. Box, G.E.P., Jenkins, G.M. and Reinsel, 1994, *Time Series Analysis*, Prentice Hall, New York (3rd Edition).

8. Brockwell, P.J. and Davis, R.A., 1996, *Introduction to Time Series and Forecasting*, Springer, New York, p. 420.

9. Chatfield, C., 1994, *The Analysis of Time Series*, Chapman and Hall (3rd Edition).

10. Jenkins, G.M. and Watts, D.G., 1968, *Spectral Analysis and its Applications*, Holden and Day, San Francisco.

11. Nagrath, I.J. and Gopal, M., 1982, *Systems Modelling and Analysis*, Tata McGraw Hill, New Delhi, p. 647.

12. Priestley, M.B., 1988, *Non-linear and Non-stationary Time Series Analysis*, Academic Press, London.

13. Raiker, P.S., 1982, *Geochemical and Stochastic Modelling of Iron Ores of Northern Goa*, PhD, IIT, Bombay.

14. Raiker, P.S. and Sahu, B.K., 1985, Single/Multiple Input – Single Output Transfer Function with Noise Models for Constituents of iron ores of Northern Goa, India, *Mathematical Geology*, 17: pp. 755-767.

15. Ripley, B.D., 1981, *Spatial Statistics*, J. Wiley, New York, p. 252.

16. Rosenblatt, M., 1985, *Stationary Sequences and Random Fields*, Birkhanger, Boston.

17. Sahu, B.K., 1983, *Proceedings of Geomodelling Workshop*, IIT, Bombay, p. 274.

18. Sahu, B.K. and Raiker, P.S., 1985, Univariate and Multivariate Stochastic Modelling of Chemical Compositions of Iron Ores in Northern Goa, India, *Mathematical Geology*, 17: pp. 317-325.

19. Shumway, R.H. and Stoffer, D.S., 2000, *Time Series Analysis and its Applications*, Springer, New York, p. 549.

20. Whitley, P., 1963, *Prediction and Regulation*, London, E.U.P.

21. Whittle, P., 1954, On stationary processes in the plane, *Biometrica*, 41: pp 434-449.

STATIONARY UNIVARIATE TIME SERIES MODELS

2.1 INTRODUCTION

Earth Sciences data are often collected in three-dimensional space as well as with reference to time at a given location. These data may be continuous along the parameter space (i.e. spatial or time dimension) but recorded/observed data are often discretized for convenience in computerized processing. The observed geological variables are called state-space, which may be either continuous or discrete random variables. Often, studies in three-dimensions can be reduced to a study in two or one-dimensional space for convenience by projection or intersection with the lower space.

Environmental changes form a four-dimensional (spatio-temporal) process, whereas mining activities usually form a three-dimensional spatial process, map studies are two-dimensional spatial process and studies along coastline, river, geological transect form a one-dimensional spatial process. Geological processes can be classified as static (without change with time) or dynamic (change with time). Long-term dynamic processes are often termed evolutionary. Short-term processes can, therefore, be treated as univariate, bivariate or trivariate time-series (although time has a 'sense' whereas space does not). Here, we consider the time series and spatial series to be mathematically equivalent and as such, the parameter space includes time as well as spatial dimension(s).

A time series is a set of observations x_t, each one being recorded at a specific time t. A discrete time series results if t and T are discrete (for example, observations made at fixed time intervals). Although earth sciences processes and data often belong to continuous parameter (time/ spatial dimensions), convenience of computer analysis requires the data input is at discrete times. The observed values at these discrete times (spatial locations) belong to state-space, which can be discrete or continuous. Often, we will assume the state-space to be a continuous random variable (modification necessary for discrete state-space is very simple: integration replaced by summation).

Time series models could be linear or non-linear and for simplicity, we deal largely with linear time series models, although many earth sciences problems are highly complex and essentially non-linear.

A linear time series has the following characteristics, some of which are found to occur jointly:

(i)   Stationarity (no level changes or trends)

(ii) Intervention (level changes may occur at discrete time(s))

(iii) Seasonality (fixed period (frequency) and amplitude of oscillation)

(iv) Oscillatory (random period (frequency) and amplitude of oscillation)

(v)   Random errors—either independent or correlated.

If so, we can forecast future values by linear combinations of a few past values, past errors or both. However, before modelling time series, we must detect any outliers (data not belonging to time series), which should be excluded from analysis.

A time series model is a suitable probability model in such a manner that each observation x_t is assumed to have realized the value of a certain random variable X_t and this concept allows for the unpredictable nature of future observations. In short, time series model specify the joint distributions (or at least means and covariances) of a sequence of random variable $\{X_t\}$, of which $\{x_t\}$ is a realization. The joint distribution of a sequence of random variables is often too large to be computed and corresponding parameters are estimated. Therefore, we often compute joint distributions up to second order (i.e. covariances) to characterize the time series model completely. It may be noted that if x_t is Normal (or if transformed x_t is Normal and modelled), then the second order cumulants are sufficient as all higher cumulants are zero. Theoretically, we should have as many realizations as possible so that the true mean value (Ensemble average) can be estimated. However, often we have only one realization of time series and may have to assume ergodicity (i.e. Time average = Ensemble average).

The simplest time series model is iid noise with no trend or seasonal components but only independent identically-distributed random errors. Therefore, $E(X_t) = 0$ and $V(X_t) = \sigma^2 > 0$. This process is useful to model more complex time series. A binary process (of coin tossing, $H = 1$ with $p = 1/2, T = -1$ with $-p = 1/2$) will have $E(X_t = 1) = 0, V(X_t) = 1$. A random walk, starting from zero, is obtained by cumulatively summing (st) iid random variables (X_t). If X_t is a binary process and $S_0 = 0$, then the random walk is a simple symmetric random walk with $E(S_t) = 0$. By differencing this random walk, we can obtain X_t as $X_t = S_t - S_{t-1}$.

We can have time series models with trends and no seasonality such as $X_t = m_t + Y_t$, where m_t is slowing varying function called trend and Y_t is a zero mean process. If m_t is a polynomial function of t, then we can use the least squares procedure to estimate m_t and then obtain $Y_t = X_t - m_t$. Other methods include either smoothing or differencing $(\nabla X_t = X_t - X_{t-1})$ $= (1 - B)X_t$ to stationary, which is a better method (less parameters and permits variable trends).

Suppose the time series has a seasonal component $s_t = s_{t-d}$ with period and a random component (Y_t). We can model s_t as a sum of harmonics

$$s_t = a_0 + \sum_{j=1}^{k} (a_j \cos \lambda_j t + b_j \sin \lambda_j t)$$

where $a_0, a_1,..., a_k, b_1,...,b_k$ are unknown parameters and $\lambda_1,..., \lambda_k$ are fixed frequencies, each being integer multiples of $2\pi/d$. We generally make $k \leq 4$ and then $j = \dfrac{2\pi m_1}{n}$ to $\dfrac{2\pi m_4}{\pi}$. The period is given by n/m_1 to n/m_4.

Strict and weak stationarity

If $\{X_t\}$ be a strictly stationary time series with $EX_t^2 < \infty$, its mean function is $E(X_t) = \mu_x(t)$ and covariance function is

$$\gamma_X(r, s) = \text{Cov}(X_r, X_s) = E(X_r - \mu_X(r)) (X_s - \mu_X(s))$$

for all integers r, s.

If $\{X_t\}$ is weakly stationary with $EX_t^2 < \infty$, its mean for $E(X_t) = \mu_X(t)$ is independent of t and

$$\gamma_{X_{(r,s)}} = \gamma_X(|r - s|)$$

which is independent of r and s.

Thus,

$$\gamma_X(h) = \gamma_X(h, 0) = \gamma_X(t + h, t)$$

which is called autocovariance function and h is termed lag. The autocorrelation function of a weakly stationary series is given by

$$\rho_X(h) = \gamma_X(h)/\gamma_X(0) = \text{Cor}(X_{t+h}, X_t)$$

The linearity of covariances is easily checked when the variances of the processes are finite and a, b, c are real constants as follows:

$$\text{Cov}(aX + bY + c, Z) = a\,\text{Cov}(X, Z) + b\,\text{Cov}(Y, Z)$$

Example 1
Verify the following:

IID noise is stationary whenever:

(i) $\sigma_2 < \infty$ as X_t are independent and identically distributed.

(ii) But white noise (WN) process with mean zero and finite variance $\sigma^2 < \infty$ is stationary, as X_t are uncorrelated but it is not necessarily the same as IID noise, although each IID noise is white noise.

(iii) A random walk S_t has $ES_t = 0$, $E(S_t^2) = t\sigma^2 < \infty$ for all t and for all $h \geq 0$

$$\gamma_s(t + h, t) = \text{Cov}(s_t, s_t) = t\sigma^2$$

depends on t, hence s_t is not stationary.

(iv) A first-order moving average process MA(1) is defined as
$X_t = z_t + \theta z_{t-1}$; $t = 0, \pm 1, ...$; θ = real constant
where $\{z_t\} = \text{WN}(0, \sigma^2)$.
We get $EX_t = 0$, $EX_t^2 = \sigma^2(1 + \theta^2) < \infty$ and

$$\gamma_X(t + h, t) = \begin{cases} \sigma^2(1 + \theta^2) & \text{if } h = 0 \\ \sigma^2 & \text{if } h = \pm 1 \\ 0 & \text{if } |h| > 1 \end{cases}$$

Thus, $\{X_t\}$ is stationary and its autocorrelation function is

$$\rho_X(h) = \begin{cases} 1 & \text{if } h = 0 \\ \theta/(1 + \theta^2) & \text{if } h = \pm 1 \\ 0 & \text{if } |h| > 1 \end{cases}$$

(v) A first-order autoregressive process AR(1) is defined

$$X_t = \phi X_{t-1} + z_t, t = 0, \pm 1, ...$$

where $\{z_t\} \sim WN(0, \sigma^2)$, $|\phi| < 1$ and z_t is uncorrelated with X_s for each $s < t$.

We easily find that $EX_t = 0$ and $\gamma_X(h) = \phi^h \gamma_X(0)$.

We then get $\rho_X(h) = \gamma_X(h)/\gamma_X(0) = \phi^{|h|}$ for $h = 0, \pm 1,\dots$.

Since z_t is uncorrelated with X_{t-1}, for linearity property, we have

$$\text{Cov}(X_t, Z_t) = \text{Cov}(\phi X_{t-1} + z_t, z_t)$$

$$= \text{Var}(z_t)$$

$$= \sigma^2$$

Hence, $\gamma_X(0) = [\sigma^2/(1 - \phi^2)] < \infty$.

From these examples of time series models, it is obvious that the autocorrelation function $\rho_X(h)$ of stationary series plays an important role in identifying the model and estimating the parameters. Later, we will also use tools like partial autocorrelation function and the independence of residual series.

Smoothing techniques

We can estimate a trend, approximately, by passing low-pass filter (q is small) with weights

$$a_j = (2q + 1) - 1$$

where $-q < j < q$.

This filter removes the rapidly-fluctuating (high frequency) component in the stationary random sequence of original data and estimates the smoothly varying trend $\{\hat{m}_t\}$.

Thus, $\hat{m}_t = \dfrac{1}{(2q+1)} \sum_{j=-q}^{q} X_{t-j}$; $q + 1 \leq t \leq n - q$.

However, if q is very large, it will reduce noise as well as allow linear trend function to pass without distortion. Therefore, we use $q = 1, 2$, or 3 corresponding to $3, 5, 7$ point smoothing functions.

For any fixed $a \in [0, 1]$, we can use one-sided moving averages (exponential smoothing) $\hat{m}_t$, $t = 1, 2, \dots, n$ as $\hat{m}_t = aX_t + (1 - a)\hat{m}_{t-1}$, $t = 2, \dots, n$ and $\hat{m}_1 = X_1$.

Thus, for $t \geq 2$, $\hat{m}_t = \sum_{j=0}^{t-2} a(1 - a)^j X_{t-j} + (1 - a)^{t-1} X_1$, a weighted moving average of $X_t, X_{t-1}, \dots$, with weight decreasing exponentially.

Classical decomposition of time series (with trend and seasonality)

$$X_t = m_t + s_t + Y_t, t = 1, ..., n$$

where $EY_t = 0$, $s_{t+d} = s_t$ (seasonality $= d$) $\sum_{j=1}^{d} s_j = 0$

If $d =$ odd, we use simple moving average with q given by $q = (d+1)/2$.

If d is even (say $2q$), we use

$$\hat{m}_t = (\tfrac{1}{2}x_{t-q} + x_{t-q+1} + \cdots + x_{t-q-1} + \tfrac{1}{2}x_{t-q})/d, d < t \leq n - q$$

The seasonal output is estimated as

$$\hat{s}_k = w_k - d^{-1}\sum_{j=1}^{d} w_t, k = 1..., d$$

and $\hat{s}_k = \hat{s}_{k-d}, k > d$.

The deseasonalized series is $d_t = x_t - \hat{s}_t$ and noise series is $Y_t = x_t - \hat{m}_t - \hat{s}_t, t = 1, ..., n$.

Alternatively, we can use the procedure of differencing, i.e. $\nabla = 1 - B$, where $BX_t = X_{t-1}$.

The seasonality of period d is obtained by writing the log-d differencing operator ∇_d as

$$\nabla_d X_t = X_t - X_{t-d} = (1 - B^d)X_t$$

We obtain $\nabla_d X_t = m_t - m_{t-d} + Y_t - Y_{t-d}$.

The trend $(m_t - m_{t-d})$ can be eliminated by using the process of differencing operator (∇), r times, making the deseasonalized series stationary. However, we must avoid overdifferencing the series (i.e. r should be as low as feasible).

Testing the residual noise series

If seasonality and trend are removed, the residual is a stationary series (noise), which should be independent and identically distributed. If $Y_1,..., Y_n$ are iid with $EY_t = 0$, $EY_t = \sigma^2$, then the sample autocorrelation for large n is $N(0,1/n)$. So, 95% of autocorrelations must lie within $\pm 1.96/\sqrt{n}$. A single portmanteau test is possible as

$$Q_{LB} = n(n + 2)\sum_{j=1}^{h} \hat{\rho}^2\,(j)/(n - j)$$

has a chi-square distribution with h degrees of freedom. We can also plot a graph of order statistics and their standard deviations, which should be linear if residuals are Normal $(Q - Q$ plot). The correlation between m_i, where $m_i = [E(Y(t)) - \mu]/r$ and $Y(t)$ should be near 1 if residuals are Normally distributed. We can also use other standard univariate tests for Normality, such as Kolmogorov-Smirnov Test, Sign Test, Turning Point Test, etc.

2.2 STATIONARY PROCESSES

We are interested in those processes whose properties (or some of their properties) do not vary with time, so that future predictions can be possible. In extrapolating deterministic functions, we assume that either the function itself or one of its derivatives is constant and then we use this constant derivative to predict the future values of the function. In probabilistic time series, we have a random component and if this random component is stationary, then we can develop powerful techniques to forecast its future values. If X_t is not stationary and some differencing can be predicted linearly, then we use Box-Cox transforms to Normalize the data; so that the transformed sequence Y_t can be predicted linearly.

$$Y_t = \begin{cases} (X_t^\lambda - 1)/\lambda, & X_t \geq 0, \lambda > 0 \\ \ln X_t, & X_t > 0, \lambda = 0 \end{cases}$$

The autocovariance function (ACVF) of a stationary time series $\{X_t\}$ is denoted as

$$\gamma(h) = \text{Cox}(X_{t+h}, X_t), h = 0, \pm 1, \pm 2, ...$$

Then, autocorrelation function (ACF) is thus given by

$$\rho(h) = \gamma(h)/\gamma(0)$$

These formulae provide powerful measures of dependence between values of a time series at different times and play important role in predicting its future values.

Example 1

If $\{X_t\}$ is a stationary, Gaussian time series and we have observed n data $x_1,..., x_n$, what is the best predictor for X_{n+h}? We know X'_n, X'_{n+h} are jointly Normal. We use the minimum mean square error as the best predictor, then the conditional distribution of $X_{n+h} \mid X_n = x_n$ is

$$N(\mu + \rho(h)(x_n - \mu)), \sigma^2(1 - \rho(h)^2)$$

The conditional mean is $\mu + \rho(h)(X_n - \mu)$ and corresponding MSE is $\sigma^2(1 - \rho(h)^2)$.

So, minimum occurs as $|\rho(h)| \to 1$ and best predictor as $\rho(h) = \pm 1$ is $\mu \pm (X_n - \mu)$.

MSE $\to 0$.

If the time series X_t is non-normal, then the joint distribution of X_{n+h} and X_n given X_t observations up to n is not normal. Hence, the conditional expectation $E(X_{n+h} | X_h)$ is not a linear function of $(X_n - \mu)$ and calculations are more complicated. However, we can still find the best linear predictor of $E(X_{n+h} | X_h)$ instead of the best function. The best linear predictor $l(X_n)$ is of the form $aX_n + b$ and we minimize $E(X_{n+h} - aX_n - b)^2$. Thus, we obtain

$$l(X_n) = \mu + \rho(h)(X_n - \mu)$$

and corresponding MSE $= \sigma^2(1 - \rho(h))^2$.

For Gaussian processes, mean value function $m(X_n)$ is the same as $l(X_n)$, best linear predictor function, which depends only on mean and ACF of $\{X_t\}$. This fact is useful since the more complicated joint distributions and conditional expectation are not to be computed.

ACVF and ACF have the following properties:

(i) $\gamma(0) \geq 0$

(ii) $|\gamma(h)| \leq \gamma(0)$ for all h

(iii) $\gamma(.)$ is even, i.e. $\gamma(h) = \gamma(-h)$ for all h.

ACF is also non-negative definite, which is defined by a function κ such that

$$\sum_{i,j=1}^{n} a_i \kappa(i - j) a_j \geq 0$$

for all positive integers n and vectors $a = (a_1, ..., a_n)'$ with real-valued components a_i.

Example 2

MA(1) process is defined as

$$X_t = z_t + \theta z_{t-1}, t = 0, \pm 1, \pm ....$$

where $\{z_t\}$ is WN($0, \sigma^2$) and θ is a real-valued constant.

So, $EX_t = 0, EX_t^2 = V(X_t) = \sigma^2(1 + \theta^2) < \infty$ (weakly stationary)

and $\gamma_X(t + h, t) = \begin{cases} \sigma^2(1 + \theta^2) & \text{if } h = 0 \\ \sigma^2 \theta & \text{if } h = \pm 1 \\ 0 & \text{if } |h| > 1 \end{cases}$

And
$$\rho_X(h) = \begin{cases} 1 & \text{if } h = 0 \\ \theta^2/(1+\theta^2) & \text{if } h = \pm 1 \\ 0 & \text{if } |h| > 1 \end{cases}$$

Real-valued solution for $\theta = (2\rho)^{-1}\left(1 \pm \sqrt{1-4\rho^2}\right)$ provided $|\rho| \le 1/2$ and then $\sigma^2 = (1 + \theta^2)^{-1}$. It may be noted that weak stationarity does not imply strict stationarity (same joint distribution function).

Example 3

AR (1) process is defined as

$$X_t = \phi X_{t-1} + z_t, \, t = 0, \pm 1, \dots.$$

where $\{z_t\}$ is WN$(0, \sigma^2)$, $|\phi| < 1$, z_t uncorrelated with X_s for each $s < t$.

We find the $EX_t = 0$ and $\gamma_X(h) = \phi\gamma_X(h-1) = \phi^h\gamma_X(0)$.

Thus, $\rho_X(h) = \gamma_X(h)/\gamma_X(0) = \phi^{|h|}, h = 0, \pm 1, \pm \dots$.

From linearity property

$$\gamma_X(0) = \sigma^2/(1-\phi^2)$$

which is real iff $|\phi| < 1$.

Example 4

MA(q) process is defined as $X_t = z_t + \theta_1 z_{t-1} + \cdots + \theta_q z_{t-q}$; θ_i = real constants where $\{z_t\}$ is WN$(0, \sigma^2)$.

Therefore, $EX_t = 0$

and
$$EX_t^2 = V(X_t) = (1 + \theta_1^2 + \cdots + \theta_q^2)\sigma^2 < \infty$$

MA(q) is always stationary and also conversely, every stationary process q-correlated time series with mean 0, then it can be represented as MA(q) process.

If $q = \infty$, the linear decomposition is termed Wold's decomposition.

Linear processes

The class of linear time series models, which include ARMA models, form a basis for studying stationary processes. Every second-order process is either a linear process or can be transformed to a linear process by subtracting a deterministic trend component. Mathematically speaking, a linear process is defined as

$$X_t = \sum_{j=-\infty}^{\infty} \psi_j z_{t-j} = \sum_{j=-\infty}^{\infty} \psi_j B^j$$

for all t, where $\{z_t\}$ is WN $(0,\ \sigma^2)$ and $\{\psi_j\}$ is a square of constants with $\sum_{j=-\infty}^{\infty} |\psi_j| < \infty$ and B = backshift operator.

A linear process is called a moving average or MA (∞) if $\psi_j = 0$ for all $j < 0$, i.e. if $X_t = \sum_{j=-\infty}^{\infty} \psi_j z_{t-j}$.

Fractionally-integrated ARMA processes belong to a more general class of linear processes for which $\sum_{j=-\infty}^{\infty} |\psi_j|$ is not summable. Operator $\psi(B)$ is a linear filter which, when applied to white noise input series $\{z_t\}$, produces the output $\{X_t\}$.

Since $\{z_t\}$ is stationary, the output $\{X_t\}$ is also stationary. Obviously, if two linear filters are applied successively, then the resulting output is stationary and it is immaterial in which order these two linear filters are applied.

Example 5

Causal stationary AR (1) process is given by

$$X_t - \phi X_{t-1} = z_t$$

where $|\phi| < 1$ and $\{z_t\}$ is WN $(0,\ \sigma^2)$ and $\{z_t\}$ is uncorrelated with X_s for each $s < t$.

We can show that

$$(1 - \phi B)X_t = z_t$$

or $X_t = (1 + \phi B + \phi^2 B^2 + ...)z_t = \pi(B)z_t = \sum_{j=0}^{\infty} \phi^j z_{t-j}$

or $\psi(B) = \phi(B)$

$$z_t = \text{MA}(\infty)$$

Example 6   *Wold decomposition*

Consider the stationary process

$$X_t = A \cos(\omega t) + B \sin(\omega t)$$

where $\omega \in (0,\ \pi)$ is constant and A, B are uncorrelated random variables with mean 0 and variance σ^2. Therefore,

$$X_n = (2 \cos \omega)X_{n-1} - X_{n-2} = \tilde{P}_{n-1}X_n,\ n = 0, \pm 1, ...$$

so that $X_n - \tilde{P}_{n-1}X_n = 0$ for all n (deterministic process).

If $\{X_t\}$ is non-deterministic, then

$$X_t = \sum_{j=0}^{\infty} \psi_j z_{t-j} + V_t$$

where $\psi_0 = 1$, $\sum_{j=0}^{\infty} \psi_j^2$, $\{z_t\} \sim (0, \sigma^2)$, $\text{Cov} \{z_s, V_t\} = 0$ for all s, t, $\{V_t\}$ is deterministic, z_t is limit to linear combinations of X_s, $s \leq t$.

If $P_t X_t$ is best linear prediction of X_t, then one-step mean square error is

$$\sigma^2 = E(X_t - \tilde{P}_{t-1} X_t)$$

The process is deterministic if $\sigma^2 = 0$.

If $\sigma^2 > 0$, the decomposition holds with one-step prediction error $z_t = X_t - \tilde{P}_{t-1} X_t$.

The term $\{V_t\}$ is perfectly predictable for process $\{X_t\}$, i.e. $\tilde{P}_t V_s = V_s$ for all s and t.

For an ARMA process, the deterministic output $V_t = 0$ for all t and such a process is termed purely non-deterministic.

A stationary time series has properties that are invariant over the time. Although some time series are not stationary, we convert these to stationary series by suitable differencing ($d > 0$ but usually < 2) of the scalar random variables. The non-stationarity can be brought back later by appropriate back-transformation, i.e. by integrating/summing d times. If the scalar variable is non-normally distributed, then we can transform it by suitable logarithmic/power transformation so that the transformed scalar variable is Normal (Gaussian) for modelling purposes and this makes the models become linear, which has mathematical advantages. For example, the sum of linear variables remains linear and if these scalars have Gaussian distributions then the sum has also Gaussian distribution.

The basic properties of stationary time series are due to its weakly stationary properties which may be reflected by constancy of covariance function $\gamma(h)$ over initial time $t = 0$ but dependence on time difference or $\log (|h| = |t - s|)$ and thus, for autocorrelation function $\rho(h) = \gamma(h)/\gamma(0)$, where $0 < \gamma(0) < \infty$. The acf $\rho(h)$ provide a measure of dependence for adjacent samples and hence are very useful for modelling and forecasting. Autocovariances and autocorrelations have following basic properties:

1. $\gamma(0) \geq 0$ indicating X_t is a random variable which, in limit, may become a constant. In addition, this also gives $\rho(0) = 1$.

2. $|\gamma(h) \le \gamma(0)|$ for all h, indicating a decrease in covariance function $\gamma(h)$ from the variance $\gamma(0)$.

3. $\gamma(h) = \gamma(-h)$ for all h, indicating it is an even function.

4. $\sum_{i,j=1}^{n} a_i \, \gamma(i-j) \, a_j \ge 0$, i.e. $\gamma(h)$ is non-negative definite for all a_i's.

If $\{X_t\}$ is strictly stationary, then X_t is identically distributed and the joint distributions for any lag h are same and independent of t. An iid sequence is strictly stationary. Every second order stationary process is either a linear process [ARMA (p, q)type] or can be transformed to it by subtracting its deterministic components. A linear process can be regarded as:

$$X_t = \sum_{j=-\infty}^{\infty} \psi_j z_{t-j}$$

$$= \psi(B)z_t$$

for all t, where $\{z_t\}$ is WN$(0, \sigma^2)$ $\{\psi\}$ is a sequence of constants with $\sum_{j=-\infty}^{\infty} |\psi_j| < \infty$. Therefore, a linear process can be called an MA(∞) process if $\psi_j = 0$ for all $j < 0$, or $X_t = \sum_{j=0}^{\infty} \psi_j z_{t-j} \equiv$ MA(∞).

The operator $\psi(B)$ is a linear filter, which operates on white noise input $\{z_t\}$ to produce stationary output $\{X_t\}$.

Let $\{y_t\}$ be a stationary time series with mean zero, covariance function y_Y and $\sum_{j=-\infty}^{\infty} |\psi_j| < \infty$.

Then the time series

$$X_t = \sum_{j=-\infty}^{\infty} \psi_j \, Y_{t-j}$$

$$= \psi(B)Y_t$$

is stationary with mean 0 and autocovariance function

$$\gamma_X(h) = \sum_{j=-\infty}^{\infty} \sum_{k=-\infty}^{\infty} \psi_j \psi_k \gamma_Y (h+k-j)$$

If X_t is also a linear process, then $\gamma_X(h) = \sum_{j=-\infty}^{\infty} \psi_j \psi_{j+h} \sigma^2$.

If we apply two linear filters $\alpha(B)$ and $\beta(B)$, then the output

$$\psi(B) = \alpha(B) \cdot \beta(B)$$
$$= \beta(B) \cdot \alpha(B)$$

i.e. the order of applying the linear filters is immaterial.

Autoregressive Models

An autoregressive process of order p, AR (p), is

$$x_t = \phi_1 x_{t-1} + \cdots + \phi_p x_{t-p} + z_t$$

where the current value is expressed as the weighted sum (multiple regression) of past p values plus current values of white noise.

Using back-shift operator B, we can briefly write it as

$$\phi_p(B)x_t = z_t$$

where the polynomial $= \phi_p(B) = 1 - \phi_1 B - \cdots - \phi_p B^p$ is termed AR(p) operator, which has p roots.

For an AR (1) process, we have

$$X_t - \phi_1 X_{t-1} = z_t$$

where $\{z_t\} \sim WN(0, \sigma^2)$, $|\phi| < 1$ and z_t is uncorrelated with X_s for each $s < t$. Taking expectation, we get $\mu = \phi\mu$ which shows mean $\mu = 0$.

The stationarity condition is $|\phi| < 1$ and easily checked by obtaining the variance as

$$\sigma_X^2 = \phi^2 \sigma_X^2 + \sigma^2 \quad \text{or} \quad (1 - \phi^2)\sigma_X^2 = \sigma^2$$

since $0 < \sigma_X^2 < \infty$ and $\sigma^2 > 0$.

This stationarity condition can be written as $(1 - \phi B)$ has its zero (root) of magnitude > 0 and for AR (p), the zeros of $\phi_p(B)$ should lie outside the unit circle.

Multiplying x_{t-h} on both sides and taking expectation, we obtain for $h \geq 1$, the autocorrelation function $\gamma_h = \phi\gamma_{h-1}$ with a solution $\gamma_h = \phi^h\gamma_0 = \dfrac{\phi^h \sigma^2}{(1 - \phi^2)}$.

This gives $\rho_h = \phi^h$ for all $h \geq 1$.

If ϕ is positive, ρ_h decays geometrically as h increases and if ϕ is negative, then ρ_h shows violent oscillations with geometric decay on both sides of $\rho_h = 0$.

The pacf $\phi_{11} = \rho_1 = \phi$ and $\phi_{hh} = 0$ for $h > 1$ (see Fig. 2.7).

The variance of sample autocorrelations for an AR (1) process as $h \to \infty$ is given by

$$\lim_{h \to \infty} \text{var}[h] \cong \frac{1}{N}\left(1 + 2\sum_{i=1}^{\infty} \phi^{2i}\right)$$

$$= \frac{1}{N}\left(\frac{1+\phi^2}{1-\phi^2}\right)$$

AR (2) process is given by

$$x_t = \phi_1 x_{t-1} + \phi_2 x_{t-2} + z_t$$

which has a mean $\mu = 0$ for stationarity, since $\phi(1) = (1 - \phi_1) \neq 0$.

By multiplying x_{t-h} and taking expectations, we obtain

for $h = 0$: $\sigma_x^2 = \phi_1 \gamma_1 + \phi_2 \gamma_2 + \sigma^2$

or $\sigma_x^2(1 - \phi_1 \rho_1 - \phi_2 \rho_2) = \sigma^2$

for $h \geq 1$: $\gamma_h = \phi \gamma_{h-1} + \phi_2 \gamma_{h-2}$

or $\rho_h = \phi_1 \rho_{h-1} + \phi_2 \rho_{h-2}$

a second order difference equation, which is easily solved.

We know, $\rho_0 = 1$ and then,

$$\rho_1 = \phi_1 + \phi_2 \rho_{-1} \text{ or } \rho_1 = \phi_1/(1 - \phi_2) \quad [\text{since } \rho_{-1} = \rho_1]$$

Then we obtain $\rho_2 = \dfrac{\phi_1^2}{1-\phi_2} - \phi_2$ by putting $h = 2$ in the difference equation.

Therefore, we get

$$\sigma_x^2\left[1 - \frac{\phi_1^2}{1-\phi_2} - \phi_2\left(\frac{\phi_1}{1-\phi_2} + \phi_2\right)\right] = \sigma^2$$

or $\sigma_x^2 = (1 - \phi_2)\,\sigma^2/(1 + \phi_2)\,(1 - \phi_1 - \phi_2)\,(1 + \phi_1 - \phi_2)$

For stationarity of the AR(2) process, each factor in the denominator of σ_x^2 formula has to be positive. Then we obtain the following conditions:

$$-1 < \phi_2, (\phi_1 + \phi_2) < 1, (-\phi_1 + \phi_2) < 1$$

Therefore, we get $\phi_2 < 1$.

Computing the pacf values, we obtain

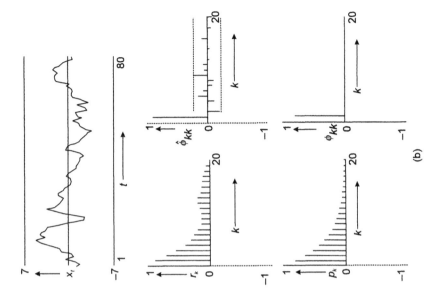

(a)

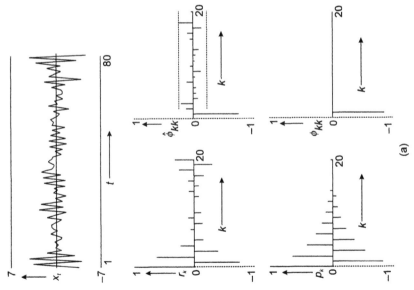

(b)

Fig. 2.1 Simulated AR(1) processes

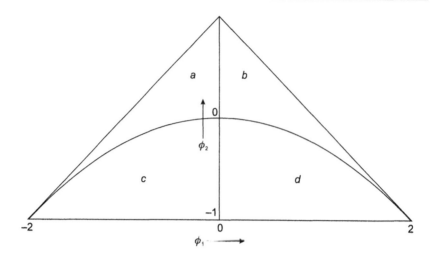

Fig. 2.2 Triangular stationary region for AR(2) processes

$$\phi_{11} = \rho_1 = \phi_1/(1 - \phi_2)$$
$$\phi_{22} = (\rho_2 - \rho_1^2)/(1 - \rho_1^2) = \phi_2$$

and $\phi_{hh} = 0$ for all $h > 2$.

For stationarity, we have $- 2 \leq \phi_1 \leq 2, - 1 < \phi_2 < 1$ which forms a triangular region with parabola through $\phi_2 = 0$ (Figs. 2.1, 2.2, 2.3, 2.4).

In the triangular stationarity diagram, we have four partitioned regions a, b, c, d defined by inequalities for stationarity conditions on ϕ_1 and ϕ_2 given earlier. In regions (c) and (d), the zeros of $\phi_2(B)$ are complex, offering the pseudo-periodic behaviour in (d) since two complex conjugate roots are multiplied to produce a real number ($i^2 = - 1$).

Example 7

Solve $\rho_h = \phi_1\rho_{h-1} + \phi_2\rho_{h-2}$.

We try the solution as $\rho_h = \lambda^h$.

Then, $\lambda^2 - \phi_1\lambda - \phi_2 = 0$.

So, $1/\lambda$ satisfies $\phi(B) = 0$ and we get the general solution as

$$\rho_h = A_1\lambda_1^h + A_2\lambda_2^h$$

where $1/\lambda_1$ and $1/\lambda_2$ are the zeros of $\phi(B)$ and A_1 and A_2 are arbitrary constants to be determined.

Since $\rho_0 = 1$, we get

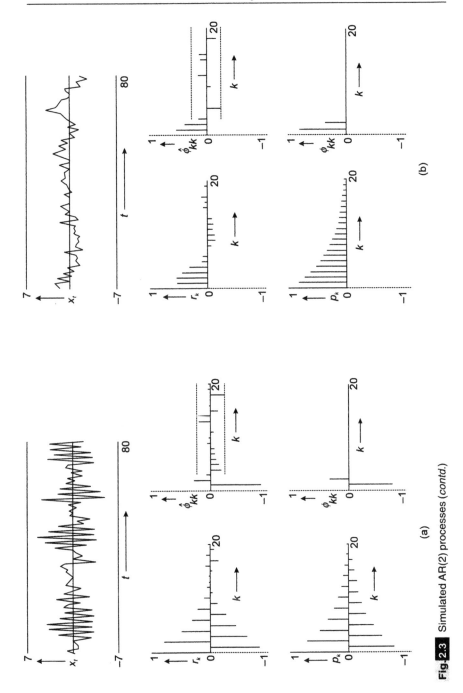

Fig. 2.3 Simulated AR(2) processes (*contd.*)

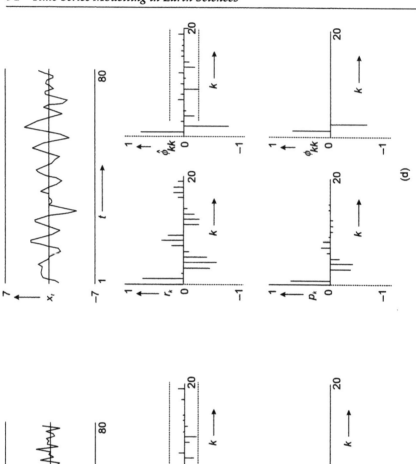

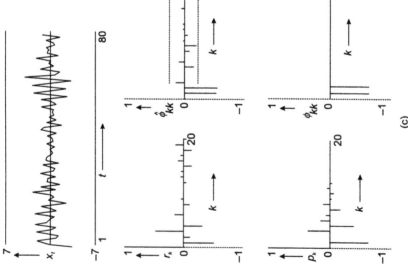

Fig. 2.4   Simulated AR(2) processes

$$A_1 + A_2 = 1$$

and
$$\rho_1 = A_1\lambda_1 + A_2\lambda_2$$

$$= \frac{1}{1-\phi_2} = \frac{\lambda_1 + \lambda_2}{1+\lambda_1\lambda_2}$$

Therefore,

$$A_1 = \frac{\lambda_1(1-\lambda_2^2)}{(\lambda_1 - \lambda_2)(1+\lambda_1\lambda_2)}$$

$$A_2 = \frac{-\lambda_2(1-\lambda_1^2)}{(\lambda_1 - \lambda_2)(1+\lambda_1\lambda_2)}$$

$E[x_i] = 0$ and recurrence for ρ_h is given by

$$\rho_h = \phi_1\rho_{h-1} + \cdots + \phi_p\rho_{h-p} . h > 0$$

The stationary values for $\rho_0, \rho_1, \ldots, \rho_{p-1}$ are obtained as was done for AR(2) process. Then, for the higher $\rho_p s$, both stationarity conditions would be more complicated but assuming stationarity, we obtain

$$\sigma_x^2(1 - \phi_1\rho_1 - \cdots - \phi_p\rho_p) = \sigma_h^2 \equiv \pi(B)\rho$$

with pacf given by

$$\phi_{hh} = \frac{|P_h^*|}{|P_h|} = \frac{\sum_{i=1}^{h} \pi_{ih}\rho_i}{|P_h|} = \phi_h$$

An alternative way of using ρ_i to write AR(p) instead of ϕ_i is

$$\phi_p(B)\rho_i = 0 \text{ for } h > 0$$

which is obtained by multiplying x_{i-h} and taking expectations for $h > 0$.

In matrix form, we can write AR (p) in Yule-Walker form as

$$\rho = P_p\phi$$

which has the solution $\phi = P_p^{-1}\phi$
where P_p^{-1} is unique since P_p is positive definite.

In actual practice, the population parameters are replaced by maximum likelihood estimators for the solution.

Example 8

Find necessary conditions for stationarity for AR (p) process.

$$x_t = \phi^{-1}_p(B)z_t$$

$$= \left[\prod_{r=1}^{p}(1 - \lambda r^B)\right]^{-1} z_i$$

where $1/\lambda_r, r = 1,....p$ are zeros of $\phi_p(B)$.

In terms of partial fractions, if each λ_r is distinct, there exist l_r such that

$$x_i = \sum_{\infty}\sum_{r=1}^{p}\frac{l_r}{(1 - \lambda_r B)}\bigg]z_i$$

Then

$$x_i = \sum_{r=1}^{p}\left(l_r\sum_{h=0}^{\infty}X_r^h\, a_{i-h}\right) = \sum_{h=0}^{\infty}\left(\sum_{r=1}^{p}l_r\lambda_r^h\right)a_{i-h}$$

and therefore, $\sigma_x^2 = \sum_{h=0}^{\infty}\left(\sum_{r=1}^{p}l_r\lambda_r^h\right)^2\sigma^2.$

For stationarity, σ_x^2 is necessarily finite, so $|\lambda_r| < r, r = 1,...,p$ that, the zeros of $\phi_p(B)$ must lie outside the unit circle. The same argument can be used for roots, which are not distinct. For example, if $\lambda_{p-1} = \lambda_p$, there exists l_r^* such that

$$x_i = \left(\sum_{r=1}^{p-1}\frac{l_r^*}{(1 - \lambda_r B)} + \frac{l_p^*}{(1 - \lambda_p B)}\right)z_i$$

and thus, $\sigma_x^2 = \sum_{h=0}^{\infty}\left(\sum_{r=1}^{p-1}l_r^*\lambda_r^h + l_p^*(h + 1)\lambda_p^h\right)^2\sigma^2$

and the same conditions are still necessary.

Example 9

Examine the form of solution when the zeros of $\phi_p(B)\rho_h = 0$ with roots $1/\lambda_r, r = 1,...,p$ are distinct and lie outside the unit circle.

We have $\phi_p(B)\rho_h = 0$ or $\rho_h = \sum_{r=1}^{p} A_r\lambda_r^h$

where $A_i, i = 1,....p$ are constants obtained by putting $\rho_0 = 1$ and solving the difference equation simultaneously for $h = 1,...(h - 1)$ we have $|\lambda_r| < 1$ for all r. There are two possibilities:

(1) Some λ_r are real, implying $A_r\lambda_r^h$ decays geometrically with/without alternate sign changes.

(2) Some pairs of λ_s, λ_t are complex conjugates with modulus λ but ρ_h are real for all h.

Therefore, A_s, A_t are also complex conjugates and

$$A_s\lambda_s^h + A_t\lambda_t^h = A\lambda^h \sin(wh + \Omega)$$

where w are Ω constants, and output is geometrically damped sine wave.

Example 10

Find the acf of process $x_i = x_{i-1} - 25x_{i-2} + z_i$.

We have $\phi_2(B) = (1 - \frac{1}{2}B)^2$ and so zeros lie outside unit circle.

This shows that the process is stationary.

The two roots are coincident and so $\rho_h = (A + C_h)\left(\frac{1}{2}\right)^h$

where $A = 1$ since $\rho_0 = 1$.

We obtain $C = 0.6$ since, $\rho_1 = \phi_1/(1 - \phi_2) = (1 + c)\frac{1}{2}$ as $\phi_1 = 1.0$ and $\phi_2 = -0.25$.

2.2.2 MA (q) Processes

Moving average process of order q is given by

$$x_i = z_i + \theta_1 z_{i-1} + \cdots + \theta_q z_{i-q}$$

where $z_i \sim WN(0, \sigma^2)$, which is abbreviated by $x_i = \theta_q(B)z_i$ with $\theta_q(B) = 1 + \theta_1 B + \cdots + \theta_q B^q$ is the MA (q) opearation.

Its mean is zero and taking variances

$$\sigma_x^2 = (1 + \theta_1^2 \cdots + \theta_q^2)\sigma^2$$

and for finite q, the process is always stationary.

If it is invertible to AR(∞) series, then

$$\theta_q^{-1}(B)\, x_i = a_i$$

or $\pi(B)x_i = a_i$

where $\pi(B) = 1 - \pi_1 B - \pi_2 B^2$.

Here, $\pi(B)$ weights must form a convergent series.

This happens if zeros of $\theta(B)$ all lie outside the unit circle, which is an analogous condition to that of stationarity condition of AR(p) series.

Similarly, a stationary AR(p) process could always be inverted to an equivalent MA(∞) one. This shows that AR and MA are equivalent and are not competing models but are of the complementary type. It would be easy to comprehend and visualize models where the number of parameters of the model, p or q, are kept as small as possible (principle of parsimony).

MA(1) Process

$$x_i = z_i (1 + \theta B)$$

where $z_i \sim WN(0, \sigma^2)$ (Fig. 2.5).

For invertibility, we have $-1 < \theta < 1$ and taking expectations, we find mean $\mu = 0$.

Also, for all lag h, $E\{x_i x_{i-h}\} = E[(z_i + \theta z_{i-1}) (z_{i-1} + \theta z_{i-1-h})]$

or $\qquad \sigma_x^2 = \gamma_0 = (1 + \theta^2) \sigma^2$

and $\qquad \gamma_1 = \theta \sigma^2, \gamma_n = 0, h > 1.$

So, acf $\rho_1 = \theta/(1 + \theta^2)$ and all higher acfs are zero. Just like AR(1) has cut-off at lag 1 for pacf ϕ_{11}, for MA(1) we have cut-off for acf at lag 1 with all higher acfs = 0.

However, the pacf for MA(1) process do not have cutoff but has geometric decay over lag h (see Figs. 2.5, 2.7).

Writing pacf as $\phi_{11} = |P_h^*| / |P_h|$ with $\rho_1 = \rho$ as all $\rho_h = 0$ for $h > 1$, we obtain

$$|P_h^*| = \begin{vmatrix} 1 & \rho & \cdots & \cdots & \rho \\ \rho & 1 & \rho & 0 & 0 \\ \cdots & \cdots & \cdots & \cdots & \cdots \\ \cdots & \cdots & \cdots & 1 & 0 \\ 0 & \cdots & \cdots & \rho & 0 \end{vmatrix} = (-1)^{h-1} \rho \begin{vmatrix} \rho & 1 & \rho & \cdots & 0 \\ \cdots & \rho & 1 & \rho & \cdots \\ \cdots & \cdots & \cdots & \cdots & \cdots \\ \cdots & \cdots & 0 & \rho & 1 \\ \cdots & \cdots & \cdots & \cdots & \rho \end{vmatrix}$$

$$= (-1)^{h-1} \rho^h$$

$$= (-1)^{h-1} \theta^h / (1 + \theta^2) h$$

$$|P_h| = |P_{h-1}| - \rho^2 |P_{h-2}|.$$

Let $\qquad |P_h| = \alpha^h$, then $\alpha^2 - \alpha + \rho^2 = 0$

or $\qquad \alpha = \dfrac{1 \pm \sqrt{1 - 4\rho^2}}{2} = \dfrac{1}{1 + \theta^2}$ or $\dfrac{\theta^2}{1 + \theta^2}.$

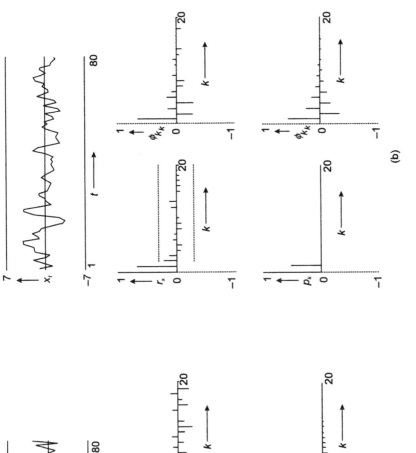

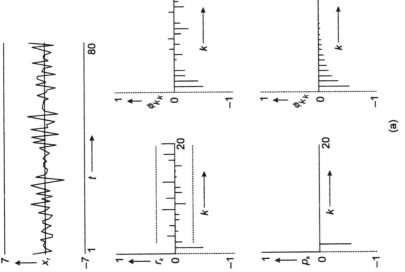

Fig. 2.5 MA(1) process

Thus, $|P_k| = (A + C\theta^{2h})/(1 + \theta^2)^h$

where $|P_1| = 1 = (1 + C\theta^2)/(1 + \theta^2)$

and $|P_2| = 1 - \rho^2 = \dfrac{1 + \theta^2 + \theta^4}{(1 + \theta^2)^2} = \dfrac{A + C\theta^4}{(1 + \theta^2)^2}$

So, $A = 1/(1 - \theta^2), C = -\theta^2/(1 - \theta^2)$

and $\phi_{hh} = |P_h^*| / |P_h| = \dfrac{(-1)^h \theta^h (1 - \theta^2)}{1 - \theta^{2(h+1)}}$

which shows that ϕ_{hh} is geometrically decreasing as h increases, using a $|\theta|$ value < 1 (since $\theta_1\theta_2 = 1$ and MA (1) process has also to be invertible) (Fig. 2.7).

MA(2) Process

$$x_i = z_i + \theta_1 z_{i-1} + \theta_2 z_{i-2}$$

where $z_i \sim WN(0, \sigma^2)$, $E(x_i) = 0$

For invertibility, by analogy with the stationarity conditions of AR(2), we obtain

$$-1 < -\theta_2$$
$$-\theta_1 - \theta_2 < 1$$
$$\theta_2 - \theta_2 < 1$$

Also, $\sigma_x^2 = \gamma_0 = (1 + \theta_1^2 + \theta_2^2)\sigma^2$
$$\gamma_1 = (\theta_1 + \theta_2\theta_1)\sigma^2$$
$$\gamma_2 = \theta_2\sigma^2$$
$$\gamma_k = 0, k > 2$$

The acfs are

$$\rho_1 = \theta_1(1 + \theta_2)/(1 + \theta_1^2 + \theta_2^2)$$
$$\rho_2 = \theta_2/(1 + \theta_1^2 + \theta_2^2)$$
$$\rho_2 = 0, k > 2$$

We see that acf has cutoff at $k > 2$ and pacf has no cutoff but slowly decays to zero.

As AR(2) had four basic patterns, MA(2) has also four patterns, corresponding to regions of invertibility as in Fig. 2.2 with ϕ_1 and ϕ_2 replaced, respectively, by $-\theta_1$ and $-\theta_2$ (see Fig. 2.6).

MA(q) Process

The stationarity is assumed to as $E[x_t] = 0$, but invertibility conditions are more complicated. The acf shows cutoffs after q lags and pacf does not show cutoff but has geometric decay to zero.

The threshold acf for MA(q) is obtained as follows (multiplying by x_{i-h} and then taking expectations):

$$\sigma_x^2 = \gamma_0 = \sum_{r=0}^{q} \theta_r^2, \, 0 < h < q$$

$$\gamma(h) = \sum_{r=0}^{q-h} \theta_r \theta_{r+h}$$

So,

$$\rho(h) = \begin{cases} \sum_{r=0}^{q-h} \theta_r \theta_{r+h} \Big/ \sum_{r=0}^{q} \theta_r^2, & 0 \le h \le q \\ 0 & \text{for } h > q \end{cases}$$

where $\theta_0 = 1$.

Example 11

Show that for any MA(q) process $|\rho_1| \le \cos\left(\dfrac{\pi}{q+2}\right)$

Ans: Consider a set of finite real MA(q) weights $\theta_0 = 1$, θ_1, ..., θ_{q-1}, $\theta_q \ne 0$ and extend the set to $\theta_{q+1} = 0$.
Then acf for MA(1) is known to be

$$\rho_1 = \sum_{r=0}^{q} \theta_r \theta_{r+1} \Big/ \sum_{r=0}^{q} \theta_r^2 \qquad \text{...(1)}$$

and we know for Cauchy-Schwartz inequality $|\rho_1| \le 1$.
Differentiating ρ_1 with respect to θ_r for $r = 1, 2, ..., q$, we have

$$\frac{\partial \rho_1}{\partial \theta_r} = \frac{\theta_{r-1} + \theta_{r+1}}{\Sigma \theta_r^2} - \frac{2\theta_r \, \rho_1}{\Sigma \theta_r^2} \qquad \text{...(2)}$$

Since MA(q) process has finite variance $\Sigma \theta_r^2 \ne 0$, therefore $\dfrac{\partial \rho_1}{\partial \theta_r} = 0$

and ρ_1 is given by

$$\theta_{r+1} - 2\rho_1 \theta_r + \theta_{r-1} = 0 \qquad \text{...(3)}$$

Since, $|\rho_1| \le 1$, we can have $|\rho_1| = \cos \alpha$ when $\dfrac{\partial \rho_1}{\partial \theta_r} = 0$ is set equal to zero.

We now differentiate $\dfrac{\partial \rho_1}{\partial \theta_r}$ with respect to θ_r again to obtain

$$\frac{\partial^2 \rho_1}{\partial \theta_r^2} = -2 \cos \alpha / \Sigma \theta_r^2 \qquad \text{...(4)}$$

which is negative and $\rho_1 = \cos \alpha$ gives maximum value when $\cos \alpha > 0$ and satisfy Eqn. (3) above.

A general solution to Eqn. (3) is

$$\theta_r = C \cos r\alpha + D \sin r\alpha \qquad \text{...(5)}$$

and using end constraints, we have

$$\theta_0 = C; 0 = C \cos(q+1)\alpha + D \sin(q+1)\alpha \qquad \text{...(6)}$$

Equation (5), therefore, reduces to

$$\theta_r = \sin(q+1-r)\alpha / \sin(q+1)\alpha$$

Substituting this value of θ_r into Eqn. (2), we obtain

$$\cos \alpha = \sum_0^q \sin(q+1-r)\alpha \sin(q-r)\alpha / \sum_0^q \sin^2(q+1-r)\alpha$$

We know that

$$\sin(q-r)\alpha = \sin(q+1-r)\alpha \cos \alpha - \cos(q+1-r)\alpha \sin \alpha$$

So, we have $0 = \sin(q+1)\alpha \sin(q+2)\alpha.$

Since $\alpha \ne k\pi/(q+1)$, for any integer k, as equality gives, $\theta_r = \infty$, which is not allowed (Eqn. 5).

We conclude that

$$\alpha = k\pi/(q+2)$$

for all integers k except multiples of $(q+2)$ which are not permitted by the relation $|\rho_1| < 1$.

The maximum value of ρ_1 is thus $\cos\left(\dfrac{\pi}{q+2}\right)$ since this is greater than zero and also greater than equal to $\cos\left(\dfrac{k\pi}{q+2}\right)$ for all values of k which are not multiples of $(q+2)$.

2.2.3 ARMA (*p*, *q*) Processes

An ARMA(p, q) process has AR operator ϕ_p (B) on x's and MA operator θ_r (B) on random shocks z_i.

We have $\qquad\qquad\qquad\quad \phi_p$ (B)$x_i = \theta_p$ (B) z_i

Stationarity and invertibility require that zeros (roots) of ϕ_p (B) and $\theta_q(B)$ must lie outside the unit circle. Taking expectation, we obtain $E[x_i] = 0$, since $\phi(1) \neq 3$.

Multiplying by $x_{i\text{-}h}$ and taking expectations, we have

$$\gamma_x (h) \equiv \gamma_h = \phi\gamma_{h-1} + \cdots + \phi_p\gamma_{h-p} + \gamma_{zx} (h) + \theta_1\gamma_{zx} (h-1) + \cdots + \theta_q\gamma_{zx} (h-q) + \cdots$$

where $\gamma_{zx}(j)$ is cross-covariance between z and x at lag difference j and is given by

$$\gamma_{zx} (j) = Cov[z_i, x_{i-j}]$$

Since $E[z_i]$ and $E[x_{i-j}]$ are both zero, we have

$$\gamma_{zx} (j) = E[z_i, x_{i-j}] \quad \text{and} \quad \gamma_{zx} (j) = 0,$$

for all $j > 0$ since z_i are independent of previous x's.

So, for $h > q$ we have the difference equation

$$\gamma_k = \phi_1\gamma_{k-1} + \cdots + \cdots \phi_p\gamma_{k-p}$$

which involves no MA parameters.

Therefore, after q lags, the autocovariances (autocorrelations) behave as an AR(p) process. But the early γ_k will have MA(q) parameters and could be complicated. The ARMA(p, q) can be written either as

$$x_i = \psi (B)z_i \text{ (useful in forecasting/updating)}$$

or as $\qquad \pi(B)x_i = z_i$

both of which are infinite series in B.

So, we expect an infinitely decaying pacf in either MA form or AR form.

For simplicity and parsimony, we investigate in details the ARMA(1,1) model where ($p + q$) = 2.
This can be written as $(1 - \phi B)x_i = (1 + \theta B)z_i$
where $z_i \sim WN (0, \sigma^2)$.
For stationarity and invertibility, we have

$$-1 < \phi < 1, -1 < \theta < 1 \quad \text{and} \quad E[x_i] = 0 \text{ [since } \phi \neq 1]$$

For all k

$$\gamma_k = \phi\gamma_{k-1} + \phi_{zx} (k) + \theta_{zx} (k-1)$$

So,
$$\gamma_0 = \phi\gamma_1 + \sigma^2 + \theta\gamma_{zx}(-1)$$
$$\gamma_1 = \phi\gamma_0 + \theta\sigma_a^2$$

and
$$\gamma_k = \phi\gamma_{k-1}, \, k > 1.$$

We also have $\gamma_{zx}(-1) - \theta\sigma^2 = \phi\sigma^2$ by multiplying the ARMA(1,1) equation by z_{i-1} and then taking expectation. This gives $\gamma_{zx}(-1) = (\theta + \phi)\sigma^2$, which is substituted in γ_0 equation. We then have,

$$\gamma_0 = (1 + 2\theta\phi + \theta^2)\sigma^2/(1 - \phi^2)$$
$$\gamma_1 = (\phi + \theta)(1 + \theta\phi)\sigma^2/(1 - \phi^2)$$

and all higher autocovariance functions/acfs are zero.

This shows that acf decays geometrically for lag 1 like an AR(1) process with parameter ϕ. The pacf decays in magnitude, with/without alternate change in sign, for $\phi_{11} = \rho_1$.

Autoregressive Moving Average Models [ARMA(p, q)]

An ARMA process $\{x_t\}$ with ACVF $\gamma_X(.)$ can well approximate a large class of ACVF $\gamma(.)$, i.e.

$$\gamma_X(.) = \gamma(h) \text{ for } h = 0, 1,..., \kappa$$

where κ is a positive integer (see Figs. 2.8, 2.9).

In addition, ARMA processes have linear structure, which leads to greatly simplified linear prediction of one-step as also multi-step future values of the process.

X_t is an ARMA(p, q) process if $\{X_t\}$ is stationary and if for every t,

$$X_t - \phi X_{t-1} - \cdots - \phi_p X_{t-p} = z_t + \theta_1 z_{t-1} + \cdots + \theta_q z_{t-q}$$

where $z_i \sim WN(0, \sigma^2)$ and polynomials $(1 - \phi_1 z - \cdots - \phi_p z^p)(1 + \theta_1 z + \cdots + \theta_p z^p)$ and have no common factors, z is a complex number, since the roots of polynomials may be complex numbers.

If $\{X_t\}$ has some finite mean μ, it can be subtracted out for each item, so that $X_t - \mu$ is ARMA(p, q) with mean zero.

The polynomials in B can be written more concisely as:

$$\phi_p(B)X_t = \theta_q(B)z_t$$

If $p = 0$, the ARMA(p, q) reduces to MA(q) and if $q = 0$, the ARMA (p, q) reduces to AR(p).

A unique stationary solution $\{X_t\}$ exists if $\phi(z) = 1 - \phi_1 z - \cdots - \phi_p z^p \neq 0$ for all $|z| = 1$ (unit circle). The ARMA(p, q) process is causal if $\phi(z) \neq 0$ for $|z| \leq 1$, i.e. roots/zeros of AR polynomial must all be greater than 1 in absolute value.

An ARMA(p, q) process $\{X_t\}$ is causal or a causal function $\{z_t\}$ if there

exists $\{\psi_j\}$ such that $\sum_{j=0}^{\infty} |\psi_j| < \infty$ and $X_t = \sum_{j=0}^{\infty} \psi_j z_{t-j}$ for all t.

Causality is equal to the condition that $\phi(z) = 1 - \phi_1 z \cdots - \phi_p z^p \neq 0$ for all $|z| \leq 1$.

Invertibility which allows z_t to be expressed in terms of X_s, $s \leq t$ can be similarly characterized in terms of MA polynomials as follows:

An ARMA(p, q) process $\{X_t\}$ is invertible if there exists constants $\{\pi_j\}$

in such a way that $\sum_{j=0}^{\infty} |\pi_j| < \infty$ and $z_t = \sum_{j=0}^{\infty} \pi_j X_{t-j}$ for all t.

Invertibility is equivalent to the condition

$$\theta(z) = 1 + \theta_1 z + \cdots + \theta_q z^q \neq 0 \text{ for all } |z| \leq 1.$$

The sequence is detected by the relation $\psi(z) = \sum_{j=0}^{\infty} \psi_j z^j = \theta(z)/\phi(z)$

or equivalently by the following identity:

$$(1 - \phi_1 z - \cdots - \phi_p z^p \, (\psi_0 + \psi_1 z + \cdots)) = 1 + \theta_1 z + \cdots + \theta_q z^q$$

Equalling coefficients of z^j for $j = 0, 1, \ldots$ etc. we obtain the recursive relation

$$1 = \psi_0$$
$$\theta_1 = \psi_1 - \psi_0 \phi_1$$
$$\theta_2 = \psi_2 - \psi_1 \phi_1 - \psi_0 \phi_2$$

or $\quad \psi_j - \sum_{k=1}^{p} \phi_k \psi_{j-k} = \theta_j; j = 0, 1, \ldots$

Similarly, $\pi_j + \sum_{k=1}^{p} \theta_k \pi_{j-k} = -\phi_j; j = 0, 1, \ldots$

In Chapter 2, we gave direct derivation formulae for $\{\psi_j\}$ and $\{\pi_j\}$.

Example 12 *AR(2) Process*

$$X_t = .7X_{t-1} - .1X_{t-2} + z_t; \{z_t\} \sim WN\,(0, \sigma^2)$$
$$\phi(z) = (1 - .5z)(1 - .2z) \text{ and zeros are at } z = 2 \text{ and } z = 5.$$

Since these roots lie outside the unit circle, $\{X_t\}$ is a causal AR(2) process with coefficients $\{\psi_j\}$ given by

$$\psi_0 = 1, \psi_1 = .7, \psi_2 = .7^2 - .1, \psi_j = .7\psi_{j-1} - .1\psi_{j-2}; j = 2, 3 \ldots$$

We can solve ψ_j either numerically or by linear difference equations.

Example 13 *ARMA(2, 1) Process*

$$X_t - .75X_t - 1 + .5625\,X_{t-2} = z_t + 1.25z_t,\; z_t \sim WN(0,\,\sigma^2)$$

The AR polynomial is $\phi(z) = 1 - .75z + .5625z^2$ having zeros at $z = 2(1 \pm i\sqrt{3})/3$ which lies outside the unit circle.

Hence, the process is causal.

But MA polynomial $\theta(z) = 1 + 1.25z$ has a zero at $z = -.8$ (inside unit circle) and hence $\{X_t\}$ is not invertible. Causality and invertibility properties are not of $\{X_t\}$ alone but with reference to both the $\{X_t\}$ as well as $\{z_t\}$ processes.

Note: If $\{X_t\}$ is ARMA(p, q) with $\phi_p(B)X_t = \theta_q(B)z$, where $\theta(z) \neq 0$ if $|z| = 1$, then it is always possible to find polynomials $\tilde{\phi}(z)$, $\hat{\theta}(z)$ and a white noise sequence $\{W_t\}$ so that $\tilde{\phi}(B)X_t = \hat{\theta}(B)W_t$ and $\tilde{\theta}(z)$ and $\tilde{\phi}(z)$ are non-zero for $|z| = 1$.

However, if the original noise sequence $\{z_t\}$ is iid, then the new white noise sequence will not be iid unless $\{z_t\}$ is Gaussian.

Therefore, we consider only stationary and invertible ARMA(p, q) processes as was done by Box and Jenkins (1970, 1976).

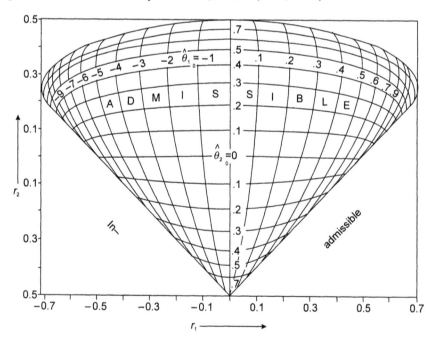

Fig. 2.6 Stationary and invertible regions in MA(2) process

ARMA (p, q) can be inverted to be represented as

$$X_t = \sum_{j=0}^{\infty} \psi_j z_{t-j}$$

where $\sum_{j=0}^{\infty} \psi_j z^j = \theta(z)/\phi(z), |z| \leq 1$

Its autocovariance function ACVF is given by

$$\gamma(h) = E(X_{t+h})X_t = \sigma^2 \sum_{j=0}^{\infty} \psi_j \psi_{j+h}.$$

For ARMA $(1, 1)$ with parameters ϕ and θ, we obtain (see Figs. 2.7-2.10)

$$\gamma_0 = \sigma^2 \left[1 + \frac{(\theta + \phi)^2}{1 - \phi^2} \right]$$

$$\gamma_1 = \sigma^2 \left[\theta + \phi + \frac{(\theta + \phi)^2 \phi}{1 - \phi^2} \right]$$

and $\qquad \gamma(h) = \phi^{h-1} \gamma(1), h \geq 2.$

An AR (q) has

$$\gamma(h) = \begin{cases} \sigma^2 \sum_{j=0}^{q-h} \theta_j \theta_{j+|h|}; & \text{if } |h| \leq q \\ 0; & \text{if } |h| > q \end{cases}$$

where $\theta_0 = 1.$
These become zero beyond q lags.

Another alternative method is to multiply X_{t-k}, K = 0, 1, 2 ... to both sides and take expectations to obtain

$$\gamma(k) - \phi_1 \gamma(k-1) \cdots - \phi_p \gamma(k-p) = 0; k \geq m$$

with $\quad m = Max(p, q+1); \psi_j = 0$ for $j < 0, \theta_0 = 1, \theta_j = 0$ for $j > q.$

The solution to this homogeneous linear difference equation is

$$\gamma(h) \cdot \alpha_1 \xi_1^{-h} + \alpha_2 \xi_2^{-h} + \cdots + \alpha_p \xi_p^{-h}, h \geq m - p$$

where $\xi_1, ... \xi_p$ are distinct roots (assumed) and $\alpha_1, ... \alpha_p$ are arbitrary constants.

We substitute this solution in the m linear difference equations and get unique values for $\alpha_1, ... \alpha_p$ and $(m - p)$ autocovariances $\gamma(h); 0 \leq h < (m - p).$

For an ARMA(1,1) process, we get

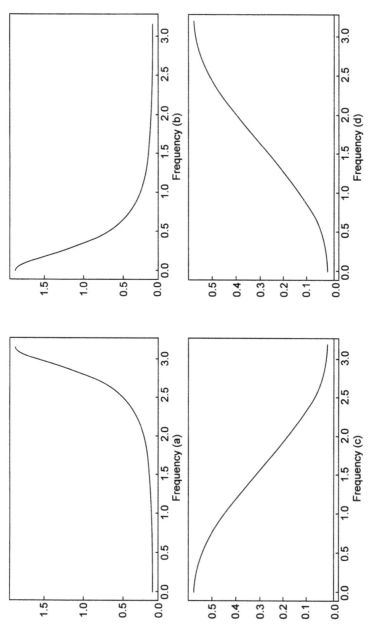

 2.7 (a) Spectral density function of AR(1) with $\rho = +\,0.7$ and error as white noise process
(b) Spectrum of MA(1) process with $\theta = +\,.7$ and error as white noise process
(c) Spectral density function of MA(1) with $\theta = +\,0.9$ and error as white noise process
(d) Spectral density MA(1) with $\theta = -\,0.9$ and error as white noise process

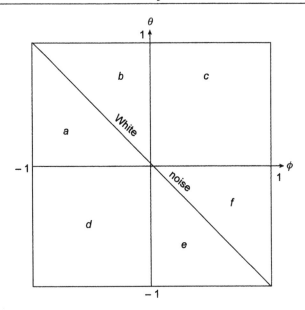

Fig. **2.8** Stationary-invertible region for ARMA(1,1) processes

$$\gamma(0) - \phi\gamma(1) = \sigma^2 (1 + \theta(\theta + \phi))$$

and $$\gamma(1) - \phi\gamma(0) = \sigma^2\theta$$

We have $\gamma(k) - \phi\gamma(k - 1) = 0, k \geq 2$ with the solution $\gamma(h) = \alpha\phi^h \ h \geq 1$.

Substituting this solution into the above two equations, we obtain

$$\gamma(0) = \sigma^2\left[1 + \frac{(\theta + \phi)^2}{1 - \phi^2}\right]; \ \gamma(1) = \sigma^2\left[\theta + \phi + \frac{(\theta + \phi)^2 \phi}{1 - \phi^2}\right]$$

and $$\gamma(h) = \phi^{h-1}\gamma(1); h \geq 2.$$

Example 14

For the causal AR(2), we define

$$(1 - \xi_1^{-1}B)(1 - \xi_2^{-1}B) X_t = z_t; \ |\xi_1|, \ |\xi_2| > 1; \ \xi_1 \neq \xi_2.$$

So, $\phi_1 = \xi_1^{-1} + \xi_2^{-1}$ and $\phi_2 = - \xi_1^{-1} \xi_2^{-1}$.

Then we get

$$\gamma(h) = \frac{\sigma^2 \xi_1^2 \xi_2^2}{(\xi_1\xi_2^{-1})(\xi_1\xi_2)}\left[(\xi_1^2 - 1)^{-1}\xi_1^{1-h} - (\xi_2^2 - 1)^{-1}\xi_2^{1-h}\right]$$

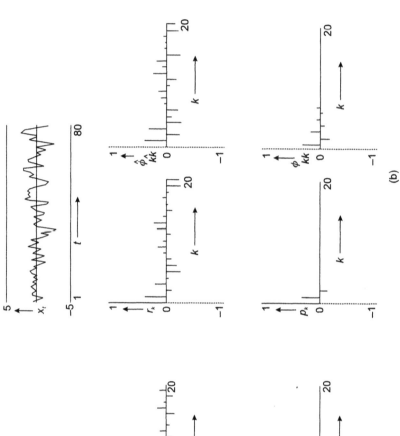

Fig. 2.9 (a), (b) Simulated ARMA(1,1) processes (*contd.*)

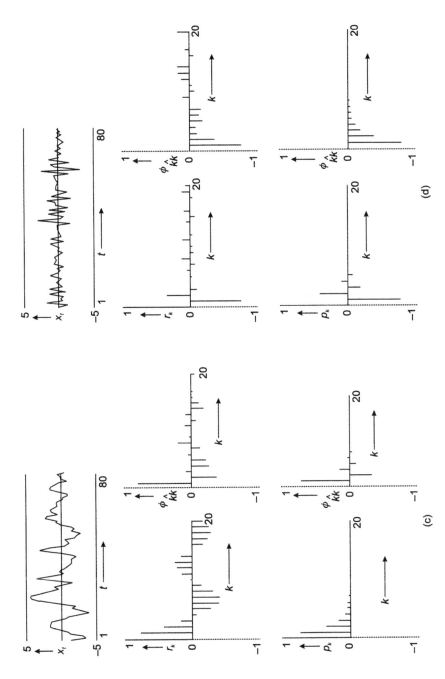

Fig. 2.9 (c), (d) Simulated ARMA(1,1) processes (*contd.*)

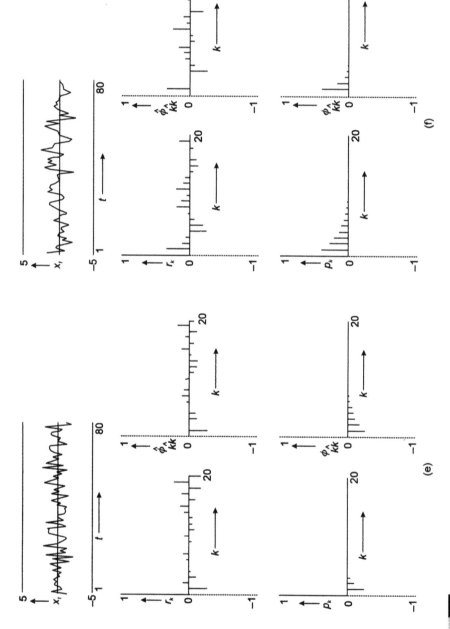

Fig. 2.9   (e), (f) Simulated ARMA(1,1) processes

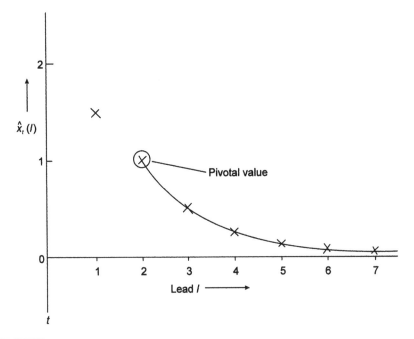

Fig. 2.10 Effective forecast function (e.f.f) for ARMA(1,2) process

If ξ_1, ξ_2 are complex conjugate roots, then we can write

$$\xi_1 = re^{i\theta}, \xi_2 = re^{-i\theta}; 0 < \theta < \pi$$

and $$\gamma(h) = \frac{\sigma^2 r^4 r^{-h} \sin(h\theta + \psi)}{(r^2 - 1)(r^4 - 2r^2 \cos 2\theta + 1)^{\frac{1}{2}} \sin \theta}$$

where $\tan \psi = \dfrac{r^2 + 1}{r^2 - 1} \tan \theta$, and $\cos \psi$ has same sign as $\cos \theta$.

This is a damped sinusoidal autocovariance function with a damping factor r^{-1} and $2\pi/\theta$.

If the roots are close to 1, the damping will be slow and we get nearly sinusoidal autocovariance function.

A third method for computing autocovariances is by solving the first $(p + 1)$ equations for $\gamma(0)$, ..., $\gamma(p)$ and then using the subsequent equations to solve successively for $\gamma(p + 1), \gamma(p + 2), ...$

This third method is convenient for numerical computations of $\gamma(h)$. With reference to ARMA(1,1) model we have:

$$\gamma(0) - \phi\gamma(1) = \sigma(1 + \theta(\theta + \phi))$$

$$-\phi\gamma(0) + \gamma(1) = \sigma^2\theta$$

Solving simultaneously for $\gamma(0)$ and $\gamma(1)$, we obtain

$$\gamma(0) = \sigma^2\left[1 + \frac{(\theta + \phi)^2}{1 - \phi^2}\right]; \ \gamma(1) = \sigma^2\left[\theta + \phi + \frac{(\theta + \phi)^2\phi}{1 - \phi^2}\right]$$

These are substituted in relation $\gamma(k) - \phi\gamma(k - 1) = 0; k \geq 2$ for obtaining $\gamma(2), \gamma(3), \dots$ etc. as $\gamma(h) = \alpha\phi^h, h \geq 1$.

ACF and PACF of stationary processes

The ACF and PACF for ARMA(1,1) can now be easily found as $\rho(h) = \gamma(h)/\gamma(0)$ for population and $\hat{\rho}(h) = \hat{\gamma}(h)/\hat{\gamma}(0)$ for sample.

In a MA(q) process, for lags beyond q, the sample autocorrelation could be independent and normal, so their autocorrelation would lie in the 95% confidence belt $\pm 1.96/\sqrt{n}$.

Partial ACF called PACF, for an ARMA process $\{X_t\}$ is a function $\alpha(.)$ defined by equations $\alpha(0) = 1$ and $\alpha(h) = \phi_{hh}, h \geq 1$, where ϕ_{hh} is the last component of $\phi_h = \Gamma_h^{-1}\gamma h$.

$$\Gamma_h = [\gamma(i - j)]_{i, j = 1}^h \quad \text{and} \quad \gamma_h = [\gamma(1), \gamma(2), \dots, \gamma(h)]'.$$

An AR(p) process is defined as
$X_t - \phi_1 X_{t-1} - \cdots \phi_p X_{t-p} = z_t, \{z_t\} \sim WN(0, \sigma^2)$ and we can get its PACF by introducing best linear predictor X_{h+1} in terms of $1, X_t, \dots, X_h$ as

$$\hat{X}_{h+1} = \phi_1 X_h + \phi_2 X_{h-1} + \phi_p X_{h+1-p}$$

Since ϕ_{hh} of X_1 is ϕ_p if $h = p$ and 0 if $h > p$, we have PACF of AR(p) as $\alpha(p) = \phi_p$ and $\alpha(p) = 0$ for $h > p$.

For $h < p$, PACF can be computed as last component of $\phi_h = \Gamma_h^{-1}\gamma_h$.
For MA(1) we have its PACF $\alpha(h) = \phi_{hh} -(-\theta)^h/(1 + \theta^2 + \cdots + \theta^{2h})$.

For AR(p), we have PACF for $h > p$ which are all zero and normally distributed with mean zero and variance $1/n$. So, the PACF should fall within the 95% confidence interval $\pm 1.96/\sqrt{n}$.

Thus, ACF and PACF are complimentary for AR(p) and MA(q) processes as regards their cut-offs in that AR(p) has PACF cutoff above p lags and MA(q) has ACF cutoff above q lags. ARMA (p, q) does not show cutoffs for ACF and PACF but behaves as AR(p) above q lags. If the modelling is correct, then the residuals could behave as a white noise $WN(0, \sigma^2)$ and the residual series can be checked for ACF and PACF of white noise series, i.e. none of the ACF and PACF of residual series is statistically significant.

2.3 IDENTIFICATION

2.3.1 Introduction

According to Box and Jenkins (1970, 1976), time series modelling follows the cycle of candidate model identification.

$$AR(p) \text{ iff } \hat{\phi}_{k\tilde{k}} N\left(0, \frac{1}{N}\right), k > q$$

$$MA(q) \text{ iff } \hat{r}_k \sim N\left(0, \frac{1}{N}\left(1 + 2\sum_1^q r_i^2\right)\right), k > q$$

ARMA(p, q) iff neither $\hat{\phi}_{kk}$ nor $\hat{r}_k$ has cut-off.

It is followed by estimation of the model parameters and verification of the candidate model (both for trend and residuals). If there is any difficulty, the model is accordingly modified and the above cycle is repeated till the candidate model is fully acceptable. In addition to graphical and visual tests, we find $E(x_t)$ and if $\mu \neq 0$, then $\hat{\mu} = \bar{x}$ is subtracted from all data to obtain $\tilde{x}_t = (x_t - \bar{x})$. We, henceforth, denote the mean corrected time series $\tilde{x}_t$ as x_t for convenience. A simple test for $E(x_t) = 0$ is to compare $\bar{x}$ with SE[$\bar{x}$] which depends upon the time series candidate model (Table 2.1 below).

Table 2.1   The variance [$\bar{x}$] for ARMA(p, q) process with $p + q \leq 2$

AR(1) $\dfrac{C_0(1 + r_1)}{N(1 - r_1)}$	ARMA(1,1) $\dfrac{C_0}{N}\left(1 + \dfrac{2r_1^2}{r_1 - r_2}\right)$	MA(1) $\dfrac{C_0(1 + 2r_1)}{(N)}$
AR(2) $\dfrac{C_0(1 + r_1)(1 - 2r_1^2 + r_2)}{N(1 - r_1)(1 - r_2)}$		MA(2) $\dfrac{C_0(1 + 2r_1 + 2r_2)}{N}$

Example 15

$$SE(\bar{x}) = \left[C_0(1 + r_1)/N(1 - r_1)\right]^{1/2} \text{ for AR (1), we have}$$

$$\sigma_{\bar{x}}^2 = \frac{1}{N} Var\left[\sum_{i=1}^{N} x_i\right] \qquad = \frac{1}{N^2}\sum_{i=1}^{N}\sum_{j=1}^{N} Cov\,(x_i, x_j)$$

$$= \frac{1}{N^2}\left(N\gamma_0 + 2\sum_{i,j>1}^{N}\gamma_{j-i}\right) = \frac{1}{N^2}\left(N\gamma_0 + 2\sum_{k=1}^{N-1}(N-k)\gamma_k\right)$$

$$= \frac{1}{N}\sum_{-\infty}^{\infty}\gamma_k \text{ (as } \gamma_k \text{ exponentially decaying)}$$

$$= \frac{1}{N}\sum_{-\infty}^{\infty}\phi^{|k|}\gamma_0 = \frac{\gamma_0}{N}\left(1 + \frac{2\phi}{1-\phi}\right)$$

$$= \frac{C_0}{N}\left(\frac{1+\hat{\phi}}{1-\hat{\phi}}\right)$$

$$\text{and SE }(\bar{x}) = \left\{\frac{C_0\,(1+r_1)}{N\,(1-r_1)}\right\}^{1/2}$$

The model parameters should be checked of admissibility of stationarity and invertibility. It is better to assume a non-stationary process even if AR parameter is close to the admissible boundary, but marginal non-stationarity may be accepted in MA(q) process.

For any MA(q) and $k \geq q$, we have

$$|\rho_k| \leq \lim_{x\uparrow q}\left(\frac{\pi}{\left[\dfrac{x+1}{k}\right]+2}\right), \text{ where } [\;] \text{ is the integer part of the enclosed part.}$$

Thus, for MA$(q)\;|\rho_1| \leq \lim_{x\uparrow q}\cos\left(\frac{\pi}{q+2}\right)$

and for MA(1) $\rho_1 \leq 1/2$ while for MA(2) $|\rho_1| \leq 1/\sqrt{2}$ and $|\rho_2| \leq 1/2$.

Thus, the ρ_k cannot be outside the admissible region for MA(q) processes. We can check whether the parameters are admissible by Table 2.2.

Table 2.2

Process	Admissible region	Initial estimate of σ^2	Initial estimates of parameters
AR(1)	$-1 < r_1 < 1$	$C_0 / \left(1 - \sum_1^p \hat{\theta}_k r_k \right)$	$\hat{\phi}_0 = r_1$
AR(2)	$-1 < r_2 < 1$		$\hat{\phi}_{10} = r_1 (1 - r_2) / (1 - r_1^2)$
	$r_1^2 < \dfrac{1}{2}(r_2 + 1)$		$\hat{\phi}_{20} (r_2 - r_1^2) / (1 - r_1^2)$
MA(1)	$-1/2 < r_1 < 1/2$	$C_0 / \left(1 + \sum_1^q \hat{\theta}_k^2 \right)$	$\hat{\theta}_0 = \left\{ 1 - \sqrt{1 - 4 r_1^2} \right\} / 2 r_1$
MA(2) see Figure 2.6 on p. 106 $\|\rho_1\| \le 1/\sqrt{2}$, $\|\rho_2\| \le 1/2$.			
ARMA	$2 r_1^2 - \|r_1\| < r_2 < \|r_1\|$	$\dfrac{C_0 (1 - \hat{\phi})^2}{(1 + 2\hat{\theta}\hat{\phi} + \hat{\theta}^2)}$	$\hat{\phi}_0 = r_2 / r_1$
(1,1)	plus or minus sign chosen such that $\|\hat{\phi}_0\| < 1$		$\hat{\phi}_0 = \left\{ b \mp \sqrt{b^2 - 4} \right\} / 2$ where $b = (1 - 2r_2 + \hat{\phi}_0^2) / (r_1 + \hat{\phi}_0)$

Example 16

(a) Show that it is impossible for ρ_1, ρ_2 for an MA(2) process to remain inside the box $\rho_1 = \pm 1/\sqrt{2}$ and $\rho_1 = \pm 1/2$ and be within the admissible region.

(b) For real valued a, b, c, we have $a^2 + b^2 + c^2 \ge -2(ab + ac + bc)$.

Thus, $-1/2 \le (ab + ac + bc) / (a^2 + b^2 + c^2)$

Choosing $a = 1, b = \theta_1$ and $c = \theta_2$ and then $a = 1, b = -\theta_1, c = \theta_2$.

$$-1/2 \le \rho_2 + \rho_1 \quad \text{or} \quad \rho_2 - \rho_1$$

Having identified the confidence model, we use a non-linear least squares procedure to estimate the vector of all the parameters

$$(\hat{\phi}, \hat{\theta}) = (\hat{\phi}_1, ..., \hat{\phi}_p, \hat{\theta}_1, ..., \hat{\theta}_q)$$

which minimizes the residual error sum of squares

$$S(\phi, \theta) = \sum_1^N \alpha_i^2$$

where $\alpha_i = \theta_q^{-1}(B)\,\phi_p(B)x_i$ are the estimated random errors given the model and the series.

Writing x_i in MA(q) form, we have

$$x_i = \hat{\alpha}_i + \hat{\pi}_1 x_{i-1} + \hat{\pi}_2 x_{i-2} + ...$$

So, $\hat{\alpha}_i = x_i - \hat{x}_i = \hat{x}_i - \overline{x}$.

We can grid search $S(\phi, \theta)$ surface over values of ϕ and θ, so that the best non-linear mle of $\hat{\phi} = (\hat{\phi}_1, ..., \hat{\phi}_p)$ and $\hat{\theta} = (\hat{\theta}_1, ..., \hat{\theta}_q)$ can be are obtained.

The approximate variances of parameters for simple models are useful in order to determine whether a parameter is significant or may be dropped for the candidate model, so that a simpler model results.

Table 2.3   Approximate variances of parameters ϕ and θ for simple models

AR(1)	Var $[\phi] = (1 - \phi^2)/N$
AR(2)	Var $[\phi_1]$, Var $[\phi_2] = (1 - \phi_2^2)/N$
MA(1)	Var $[\theta] = (1 - \theta^2)/N$
MA(2)	Var $[\theta_1]$, Var $[\theta_2] = (1 - \theta_2^2)/N$
ARMA(1,1)	$\text{Var}\left[\hat{\phi}\right] = \left(\dfrac{1 - \phi^2}{N}\right)\left(\dfrac{(1 + \phi\theta)^2}{\phi + \theta)^2}\right)$
	$\text{Var}\left[\hat{\theta}\right] = \left(\dfrac{1 - \theta^2}{N}\right)\left(\dfrac{(1 + \phi\theta)^2}{\phi + \theta)^2}\right)$

In ARMA(1,1), $\theta + \phi$ cannot be equal to zero where the AR component cancels the effect of MA or vice versa, so that ARMA(1,1) becomes redundant and is reduced to a white noise model $x_i = z_i$.

The proof is simple.

$(1 - \phi B)x_i = (1 + \phi B)z_i$

or $\qquad x_i = (1 + \phi B + \phi^2 B^2 + \cdots)(1 + \theta B)z_i$

or $\qquad x_i = (1 + \phi B)(1 + \theta B)z_i \qquad$ (when ϕ is very small).

$\qquad\qquad x_i = \{1 + (\phi + \theta)B\}z_i \qquad$ (when θ and ϕ both are small).

$\qquad\qquad x_i = z_i \qquad\qquad\qquad\qquad$ (when $\phi + \theta$ is very small).

Alternatively, $(1 - \theta B + \theta^2 B^2 - \cdots)(1 - \theta B)x_i = z_i$

$\qquad\qquad (1 - \theta B(1 - \phi B))x_i \qquad$ when θ is small

$\qquad\qquad \{1 - (\phi + \theta)B\}x_i \cong z_i \qquad$ when θ and ϕ both are small.

$\qquad\qquad x_i = z_i \qquad\qquad\qquad\qquad$ when $(\phi + \theta)$ is very small.

For an ARMA(1,1) model, a small change in θ or ϕ can be nearly compensated by a suitable change in the other parameter. Due to this duality, we may face some difficulty in estimating parameters θ or ϕ.

Once the model has been accepted and its parameters are estimated, we can check its adequacy through some diagnostic checks such as:

(i) overfitting by one parameter improves significant decrease in the model σ^2 or not;

(ii) whether residuals are having $WN(0, \sigma^2)$ distribution; and

(iii) any redundancy should be eliminated and parsimony in the number of model parameters to be attempted.

Example 17

We fit an MA(1) model and consider whether ARMA(1,1) overfit is desirable with the following values for parameter estimates:

Table 2.4

MA(1)	$\hat\theta = .845$	SE $= .038$	$\sigma^2 = .833$	$\chi^2_{residual} = 22.45$
ARMA(1,1)	$\phi = -.004$	SE $= .082$		
	$\theta = .854$	SE $= .041$	$\sigma^2 = .830$	$\chi^2_{residual} = 21.77$
MA(2)	$\theta_1 = .824$	SE $= .068$		
	$\theta_2 = -.026$	SE $= .027$	$\sigma^2 = .831$	$\chi^2_{residual} = 21.00$

The overfit does not reduce σ^2 for the residuals nor $\chi^2_{residual}$ values either for ARMA(1,1) or for MA(2).

Hence, we conclude that overfitting is not necessary and MA(1) is an acceptable model.

Example 18

A sequence of 200 data gave MA(2) model and ARMA (1,1) was also fitted as an alternative. The results are:

Table 2.5

		SE	σ^2	$\chi^2_{residual}$
MA(2)	$\hat{\theta}_1 = -1.010$	.053		
	$\hat{\theta}_2 = 0.635$	.054	.864	20.98
ARMA(1,1)	$\hat{\phi} = -.631$	.073		
	$\hat{\theta} = -.269$	.077	.997	49.82

The residual variance for MA(2) is smaller and χ^2 value is about 60% less than that for ARMA(1,1) model. Hence, we accept MA(2) model even if ρ_1 is $-.719$. The amount of overshooting is nearly .012 and

$$\text{Var } [r_1] = \frac{1+2(.7)^2}{200} = .01 \text{ with SE} = 0.1.$$ Hence, the overshooting is insignificant.

Subtracting any acceptable ARMA(p, q) model values from the corresponding data (corrected for means) will result in the residuals z_i which has distribution $WN(0, \sigma)$ with $N \to$ large. Therefore, their autocorrelations should be uncorrelated for lags $k > 5$ (say),

i.e. $\chi_k(\hat{z}) \sim WN(0, 1/N)$.

An easy but low-powered check is the portmanteau lack of fit test where we obtain the statistics

$$R = N \sum_{k=1}^{\kappa} \gamma_k^2 (z)$$

for κ equal to, say 20.

Then R has $\chi^2_{\kappa - p - q}$ distribution for testing purposes. However, this test may not be much useful in model discrimination/solution as its power is rather low.

Example 19

MA(1) fit with $\hat{\theta}$ was accepted and $\gamma_1(\hat{z}) = -.12$, $\gamma_2(\hat{z}) = .02$ with $\gamma_3(\hat{z})$ to $= -.465$ $\gamma_{20}(\hat{z})$ each < 0.2.

If MA(1) is model, then its shocks behave as AR(1) process with $x_i = .465x_{i-1} + z_i$.

So, variance of residuals Var $[\gamma_1(\hat{z})] = .465^2/100$ and SE $[\gamma_1(\hat{z})] = .047$ and $\gamma_1(\hat{z})$ is significant at about 1% level. The higher $\gamma_k(\hat{z})$ are not significant. Thus, residual shocks are not white noise but MA(1) process.

Example 20

AR(2) process was simulated to generate 200 data and fitting with various models gave the following results:

Table 2.6

		SE	σ_z^2	χ_{res}^2
AR(1)	$\hat{\phi} = .429$	.065	.926	64.18
AR(2)	$\hat{\phi}_1 = .285$	.067		
	$\hat{\phi}_2 = .360$	.067	.805	18.43
ARMA(1,1)	$\hat{\phi} = .781$	.071		
	$\hat{\theta} = -.424$	.111	.861	36.59
AR(3)	$\hat{\phi}_1 = .348$	.072		
(overfit)	$\hat{\phi}_2 = .405$	.070		
	$\hat{\phi}_3 = -.168$	.071	.785	12.89

The χ^2 for AR(1), being highly significant, is rejected. ARMA(1,1) has highly significant χ^2, so it is also rejected. AR(3) model gives lower σ_z^2 as well as lower χ_{res}^2 and may be acceptable. But AR(2) model has same σ_z^2 value and AR(3) overfit seems redundant, especially since χ_{res}^2 is non-significant for AR(2) as well at 5% level.

Another procedure is cumulative periodogram check for series $\hat{z}_1, ..., \hat{z}_N$ by

$$I(r) = \frac{2}{N}\left\{\left(\sum_{i=1}^{N} \hat{z}_i \cos 2\pi\frac{r}{N}i\right)^2 + \left(\sum_{i=1}^{N} \hat{z}_i \sin 2\pi\frac{r}{N}i\right)^2\right\}$$

for $\quad r = 1, 2, ..., \left[\dfrac{N-2}{2}\right].$

Then the standardized cumulative periodogram is

$$C(j) = \left(\sum_{r=1}^{j} Ir\right) / N \, \hat{r}_z^2 \text{ for } j = 1, 2 \dots, \left[\frac{N-2}{2}\right] \text{ with } C[N/2] = 1.$$

The plot will be scattered randomly about the straight line joining (0,0) with (1/2,1). Independent fit shows as systematic deviations for this straight line and its significance assessed Kolmogorov-Smirnoff (± 1.36) at 5% level; (± 1.63 at 1% level).

2.3.2 Parametric Estimation

The determination of an appropriate ARMA(p, q) to represent an observed stationary time series include:

   (i) choice of p and q or order selection;
  (ii) estimation of parameters ($p + q$) ($\phi_i, i = 1, \dots, p$ and $\theta_i, i = 1, \dots, q$) and error variance σ^2; and subsequently
(iii) goodness of fit tests, including AICC statistics.

If we assume p and q to be known, then we can estimate parameters $\phi_i, i = 1, \dots, p$ and $\theta_i, i = 1, \dots, q$ and corresponding error variance σ^2. We assume the mean value μ is zero; otherwise $\overline{X}$ (sample mean) is subtracted from X_t such that $(X_t - \overline{X})$ now has zero expectation. We fit ARMA(p, q) to mean corrected data as

$$\phi(B)X_t = \theta(B)z_t, \{z_t\} \sim \text{WN } (0,1/N),$$

which has $p + q + 1$ parameters and these parameters can be estimated using the maximum likelihood (Gaussian) method. The likelihood is not a quadratic function of the unknown parameters and, therefore, we have to use the non-linear method for searching the global maximum. If the initial estimates are close to the values of parameters, then the non-linear procedure will converge very fast.

Preliminary estimation for AR(p) is to use either Yule-Walker or Burg methods whereas for MA(q) processes, we use innovations or Hannan-Rissanen algorithm. After fitting the preliminary model, we search for the maximum Gaussian likelihood estimators, using the innovation algorithm. Then confidence intervals for the estimated coefficients may be easily obtained.

AR(p) Model

Burg's algorithm usually provides higher likelihoods than Yule-Walker equations. For MA(q), innovations usually yield slightly higher likelihoods than Hannan-Rissanen algorithm while for causal ARMA (p, q), the Hannan-Rissanen method is preferable.

Since $\theta(z) = 1$ and causality is assumed, we obtain

$$X_t = \sum_{j=0}^{\infty} \psi_j z_{t-j}$$

where $\qquad \psi(z) = \sum_{j=0}^{\infty} \psi_j z^j = 1/\phi(z).$

Multiplying by X_{t-j}, $j = 0,....,\ p$ and taking expectation, we obtain Yule-Walker equations

$$\Gamma_p \phi = \gamma_p$$

and $\qquad \sigma^2 = \gamma(0) - \phi' \gamma_p.$

These equations are used to determine ACVF $\gamma(0),\ ...,\ \gamma(p)$ for σ^2 and ϕ.

Since $\hat{\Gamma}_m$ is non-singular for every $m = 1, 2, ...$ we get

$$\hat{\phi} = (\hat{\phi}_1, ..., \hat{\phi}_p)' = \hat{R}_p^{-1} \hat{\rho}_p$$

and $\qquad \hat{\sigma}^2 = \hat{\gamma}(0) \left[1 - \hat{\rho}_p' \hat{R}_p^{-1} \hat{\rho}_p \right]$

where $\qquad \hat{\rho}_p = \hat{\gamma}_p / \hat{\gamma}(0).$

The large sample distribution of $\hat{\phi}$ from an AR(p) process is

$$\hat{\phi} \approx N(\phi, n^{-1}\ \sigma^2\ \Gamma_p^{-1})$$

Replacing σ^2 by $\hat{\sigma}^2$ and Γ_p by $\hat{\Gamma}_p$, we get a large sample confidence intervals for ϕ. The partial autocorrelations have $N(0,1/n)$ and 95% confidence interval is given by $\pm 1.96/\sqrt{n}$ for $k > m$ for all $k > p$. A better approach for order selection is to find values of p and ϕ_p that minimize AICC as follows:

$$\text{AICC} = -2 \ln L(\phi_p, S(\phi_p)/n) + 2\ (p+1)n/(n-p-2)$$

where L is Gaussian likelihood for ARMA process:

$$L(\phi, \theta, \sigma^2) = \frac{1}{(2\pi\sigma^2)^{n/2}(\gamma_0 - \gamma_{n-1})^{1/2}} \exp\left\{ -\frac{1}{2\sigma^2} \sum_{j=1}^{n} \frac{(X_j - \hat{X}_j)^2}{\gamma_j - 1} \right\}$$

where
$$\gamma_n = \frac{1}{\sigma^2} E(X_{n+1} - \hat{X}_{n+1})^2 = E(W_{n+1} - \hat{W}_{n+1})^2$$

We increase of order p step by step by one unit and check the resulting fit; then we can choose the best fit model.

Burg's Algorithm for AR(p)

Burg's algorithm estimates PACF $\{\phi_{11}, \phi_{22}, ...\}$ by successively minimizing sums of squares of forward and backward one-step prediction errors with respect to the coefficients ϕ_{ii}. The forward and backward prediction errors $u_i(t)$ and $v_i(t)$ satisfy recursives

$$u_0(t) = v_0(t) = x_{n-1-t}$$

$$u_i(t) = u_{i-1}(t-1) - \phi_{ii} v_{i-1}(t)$$

and
$$v_i(t) = v_{i-1}(t) - \phi_{ii} u_{i-1}(t-1)$$

Burg's estimate $\phi_{11/(B)}$ of ϕ_{11} is obtained by minimizing

$$\sigma_1^2 = \frac{1}{2(n-1)} \sum_{t=2}^{n} [u_1^2(t) + v_1^2(t)]$$

with respect to ϕ_{11}.

We obtain $u_1(t)$, $v_1(t)$ and σ_1^2 which are substituted in the above equations with $i = 2$.

Then we minimize $\sigma_2^2 = \frac{1}{2(n-2)} \sum_{t=3}^{n} [u_2^2(t) + v_2^2(t)]$ with respect to ϕ_{22} in order to obtain $\phi_{22(B)}$ of ϕ_{22} and corresponding values $u_2(t)$, $v_2(t)$ and σ_2^2.

The process is continued up to $\phi_{pp(B)}$ and get $\sigma_p^2, p \leq n - 1$.

Estimates of ϕ_{pj} in best linear predictor $P_p X_{p-1} = \phi_p X_p + \cdots + \phi_{pp} X_1$ is obtained by substituting $\phi_{ii(B)}, i = 1, ..., p$ for ϕ_{ii}.

The large sample distribution is same as Yule-Walker coefficients and is given by $N(\phi, n^{-1}\sigma^2 \Gamma_p^{-1})$.

Innovations Algorithm for MA(q)

The fitted innovations MA(m) model is given by

$$X_t = z_t + \hat{\theta}_{m1} z_{t-1} + \cdots + \hat{\theta}_{mm} z_{t-m} ; \{z_t\} \sim WN(0, \hat{v}_m)$$

If X_t is invertible MA(q) with $\{z_t\} \sim IID\ (0,\ \sigma^2)$ with $Ez_t^4 < \infty$, $\theta_0 = 1$, $\theta_j = 0$ for $j > q$.

Then innovation estimates have large sample properties.

If $n \to \infty$ and $m(n)$ is a sequence of +ve integers such that $m(n) \to \infty$ but $n^{-1/3}\ m(n) \to 0$, then for each +ve integer k, the joint distribution function of

$$n^{1/2}(\hat{\theta}_{m1} - \theta_1,\ \hat{\theta}_{m2} - \theta_2,\ ...,\ \hat{\theta}_{mk} - \theta_k)'$$

converges to MND with mean 0 and covariance matrix A in such a way that

$$a_{ij} = \sum\nolimits_{t=1}^{\min(i,j)} \theta_{i-r}\,\theta_{j-r}$$

This gives consistent estimation $\hat{v}_m$, for σ^2.

We can use order selection since for $m > q$, $\rho(m) = 0$ and variance

$$n^{-1}[1 + 2\rho^2(1) + \cdots + 2\sigma^2(q)].$$

This gives the significant values of $\rho(m)$ so that m in MA(m) is defined. In addition, we can study the values of coefficients θ_{mj} since as m is increased, these values will stabilize.

We can also use AICC statistic as

$$\text{AICC} = -2\ln \times 2\ (\theta_q,\ S(\theta_q)/n) + 2(q+1)n/(n-q-2)$$

where L is the Gaussian likelihood.

Approximate 95% confidence boundaries for θ_j are given by

$$\hat{\theta}_{mj} \pm 1.96n^{-1/2}\left(\sum\nolimits_{i=0}^{j-1} \hat{\theta}_m^2\right)^{1/2}$$

For ARMA(p, q) model, AICC is given by

$$\text{AICC} = -2\ln \times 2\ (\phi_p,\ \theta_q,\ S(\phi_p,\ \theta_q)/n) + 2(p+q+1)n/(n-p-q-2).$$

Hannan-Rissanen Algotithm for MA(q)

In this procedure, even if $q \neq 0$, we regress X_t on $X_{t-1}\ ...\ X_{t-p}$ and $\hat{z}_{t-1},\ ...,$ $\hat{z}_{t-q}$, where $\hat{z}_{t-1},\ ...,\ \hat{z}_{t-q}$ are estimated erros to obtain preliminary estimates for $\phi_1,\ ...,\ \phi_p$ and $\theta_1,\ ...\ \theta_q$.

The steps are as follows:

Step 1

A high order AR(m) (with $m >$ Max (p, q))is fitted using Yule-Walker algorithm as above and $\hat{z}_t$ is estimated as $\hat{z}_t = X_t - \hat{\phi}_{m1}X_{t-1} - \cdots - \hat{\phi}_{mm}X_{t-m}$, $t = m + 1, ..., n$.

Step 2

Using least squares we now estimate the vector of parameters, $\beta = (\phi', \theta')'$ by regression of X_t onto $(X_{t-1} \cdots X_{t-p}, \hat{z}_{t-1}, ..., \hat{z}_{t-q})$, $t = m + 1, ..., n$.

We obtain the Hannan-Rissanen estimator

$$\hat{\beta} = (z'z)^{-1}z'X_n,$$

where $X_n = (X_{m+1}, q, ..., X_n)'$ and z is $(n - m - q) \times (p + q)$ matrix given by

$$z = \begin{bmatrix} X_{m+q} & X_{m+q+1} \cdots X_{m+q+1-p} & \hat{z}_{m+q} \cdots \hat{z}_{m+1} \\ X_{m+q+1} & X_{m+q} \cdots X_{m+q+2-p} & \hat{z}_{m+q+1} \cdots \hat{z}_{m+2} \\ X_{n-1} & X_{n-2} \cdots X_{n-p} & \hat{z}_{n-1} \cdots \hat{z}_{n-q} \end{bmatrix}$$

The H-R estimate of white noise values is

$$\hat{\sigma}_{HR}^2 = S(B)/(n - m - q)$$

where $S(B)$ is sum of squares of errors with t varying from $(m + 1 + q)$ to n.

In order to improve the estimates for parameters as $\hat{z}_{t-1}, ..., \hat{z}_{t-q}$ are only estimates and not observed values, we can follow Step 3.

Step-3

Using estimate $\hat{\beta} = (\hat{\phi}_1, ..., \hat{\phi}_p, \hat{\theta}_1, ..., \hat{\theta}_p)'$ as obtained in Step 2, set

$$\tilde{z}_t = \begin{cases} 0 & \text{if } t \leq \max(p, q) \\ X_t - \sum_{j=1}^{p} \hat{\phi}_j X_{t-j} - \sum_{j=1}^{q} \hat{\theta}_j \tilde{z}_{t-j} & \text{if } t > \max(p, q) \end{cases}$$

Then for $t = 1, ..., n$, we express

$$V_t = \begin{cases} 0 & \text{if } t \leq \max(p, q) \\ \sum_{j=1}^{p} \hat{\phi}_j V_{t-j} + \tilde{z}_t & \text{if } t > \max(p, q) \end{cases}$$

$$W_t = \begin{cases} 0 & \text{if } t \leq \max(p,q) \\ -\sum_{j=1}^{p} \hat{\theta}_j W_{t-j} + \tilde{z}_t & \text{if } t > \max(p,q) \end{cases}$$

If $\hat{\beta}^*$ minimizes

$$S^*(\beta) = \sum_{t=\max(p,q)+1}^{n} \left(\tilde{z}_t - \sum_{j=1}^{p} \beta_j V_{t-j} - \sum_{k=1}^{q} \beta_{k+p} W_{t-k} \right)^2$$

then the improved estimation for β is $\hat{\beta}^* + \hat{\beta}$.

The new estimator $\tilde{\beta}$ has the same asymptotic efficiency as the maximum likelihood estimator. Another method for computing the ML estimates is to use model values of Step 2 for numerical maximization of estimator.

ML Estimation

Let $\{X_t\}$ be a Gaussian time series with zero mean and autocovariance function

$$k(i,j) = E(X_i X_j).$$

Let $\quad X_n$

$$= (X_1, ..., X_n)' \quad \text{and} \quad \hat{X}_n = (\hat{X}_1, ..., \hat{X}_n)'$$

where $\quad X_1 = 0$ and $\hat{X}_j = E(X_j + X_1 ..., X_{j-1}) = P_{j-1} X_j, j \geq 2$

Let Γ_n denote the covariance matrix $\Gamma_n = \pm (X_n X_n')$ and assume Γ_n to be non-singular.

The likelihood of X_n is

$$L(\Gamma_n) = (2\pi)^{-n/2} (\det \Gamma_n)^{-1/2} \exp\left(-\frac{1}{2} X_n' \Gamma_n^{-1} X_n \right)$$

Direct calculation of Γ_n and Γ_n^{-1} can be found by computing one-step prediction errors $X_j - \hat{X}_j$ and their variables and both these are easily calculated recursively from innovation algorithm. The one-step prediction errors are independent and their covariance matrix will be diagonal, i.e.

$$D_n = \text{diag} \begin{cases} v_j & v \\ 0 & n-1 \end{cases}$$

From innovation algorithm, we have

$$\Gamma_n = C_n D_n C'_n$$

where C_n is a lower triangular matrix such that $X_n = C_n(X_n - \hat{X}_n)$.

So, $X'_n \Gamma_n^{-1} X_n = \sum_{j=1}^{n} (X_j - \hat{X}_j)^2 / v_{j-1}$

and $\det \Gamma_n = (\det C_n)^2 (\det D_n) = v_0 v_1 \cdots v_{n-1}$.

The likelihood of vectors X_n is, therefore,

$$L(\Gamma_n) = \frac{1}{\sqrt{(2\pi)^n \, v_0 \cdots v_{n-1}}} \exp\left\{ -\frac{1}{2} \sum_{j=1}^{n} (X_j - \hat{X}_j)^2 / v_{j-1} \right\}$$

The justification for using maximum Gaussian likelihood estimators of ARMA coefficients is that large-sample distribution of the estimates is the same for $\{z_t\} \sim IID\ (0, \sigma^2)$ regardless of whether or not $\{z_t\}$ is Gaussian. The one-step procedures are obtained as

$$\hat{X}_{n+1} = \begin{cases} \sum_{j=1}^{n} \theta_{nj} (X_{n+1-j} - \hat{X}_{n+1-j}),\ 1 \le n \le m \\ \phi_1 X_n + \cdots + \phi_p X_{n+1-p} + \sum_{j=1}^{n} \theta_{nj}(X_{n+1-j} - \hat{X}_{n+1-j}),\ n \ge m \end{cases}$$

and $E(X_{n+1} - \hat{X}_{n+1})^2 = \sigma^2 E(W_{n+1} - \hat{W}_{n+1})^2 = \sigma^2 \gamma_n$,

where θ_{nj} and γ_n are obtained from innovation algorithm.

Using those values, we obtain a Gaussian likelihood estimator for an ARMA(p, q) process

$$L(\phi, \theta, \sigma^2) = \frac{1}{\sqrt{(2\pi\sigma^2)^n \, \gamma_0 \cdots \gamma_{n-1}}} \exp\left\{ -\frac{1}{2\sigma^2} \sum_{j=1}^{n} \frac{(X_j - \hat{X}_j)^2}{\gamma_{j-1}} \right\}$$

From the above procedures, we obtain parameters by differentiating and equating to zero

$$\hat{\sigma}^2 = n^{-1} S(\hat{\phi}, \hat{\theta})$$

where $S(\hat{\phi}, \hat{\theta}) \sum_{j=1}^{n} = \frac{(X_j - \hat{X}_j)^2}{\gamma_{j-1}}$

and $\hat{\phi}, \hat{\theta}$ values minimize (done numerically).

$$l(\phi, \theta) = \ln(n^{-1}S(\phi, \theta)) + n^{-1}\sum_{j=1}^{n}\ln \gamma_{j-1}$$

Least Square Estimation for ARMA (p, q) Models

We minimize $S(\tilde{\phi}, \tilde{\theta}) = \displaystyle\sum_{j=1}^{n}\frac{(X_j - \hat{X}_j)^2}{\gamma_{j-1}}$ subject to the constraints that the

model is causal and invertible.
The LS estimate of σ^2 is given by

$$\tilde{\sigma}^2 = S(\tilde{\phi}, \tilde{\theta})/(n-p-q).$$

Order Selection

We use the AICC criterion as follows:
Choose p, q, ϕ_p, θ_q to minimize

$$\text{AICC} = -2\ln L(\phi_p, \theta_q, S(\phi_p, \theta_q)/n) + \frac{2(p+q+1)n}{(n-p-q-2)}$$

If p and q are fixed, then we minimize
$$\text{AICC} = -2\ln L(\phi_p, \theta_q, \sigma^2)$$
and then parameters are the ML estimators.

Final decisions for order selection are to be made using AICC statistic (not preliminary estimators).

Confidence regions for the coefficients are given by

$$\hat{\beta} \approx N(\beta, n^{-1}V(\beta))$$

where $V(\beta)$ is the covariance matrix of β.
For special case of AR(p), we have
$$V(\phi) = \sigma^2\Gamma^{-1}$$
and, therefore, $V(\phi) = (1 - \phi_1^2)$ for AR(1)

and $\qquad V(\phi) = \begin{bmatrix} 1-\phi_2^2 & -\phi_1(1+\phi_2) \\ -\phi_1(1+\phi_2) & 1-\phi_2^2 \end{bmatrix}$ for AR(2) processes.

For MA(q) processes, we have

$$V(\theta) = \Gamma^*_q$$

Γ^*_q being the covariance matrix $Y_1, ..., Y_q$.

Here, $\{Y_t\}$ is the equivalent (replace ϕ_i with $-\theta_i$) autoregressive process with AR polynomial $\theta(z)$.

So, for MA(1), $V(\theta) = (1 - \theta_1^2)$ and for MA(2) $V(\theta)$

$$= \begin{bmatrix} 1 - \theta_2^2 & -\theta_1(1 + \theta_2) \\ +\theta_1(1 + \theta_2) & 1 - \theta_2^2 \end{bmatrix}$$

ARMA(1,1), $V(\phi, 0) = \begin{bmatrix} (1 - \phi^2)(1 + \phi\theta) & -(1 - \theta^2)(1 - \phi^2) \\ -(1 - \theta^2)(1 - \phi^2) & (1 - \phi^2)(1 + \phi\theta) \end{bmatrix}$.

Diagnostic Checking

The residuals for an ARMA(p, q) model are

$$\hat{W}_t = (X_t - \hat{X}(\hat{\phi}, \hat{\theta})) / (\gamma_{t-1}(\hat{\phi}, \hat{\theta}))^{1/2}; t = 1, 2, ..., n.$$

If ARMA(p, q) is a true model, the $\{\hat{W}_t\} \sim WN(0, \hat{\sigma}^2)$.

However, if p, q are not known (can change), then $\{\hat{W}_t\}$ is not white noise but has properties similar to those of WN sequence.

$$W_t(\phi, \theta) = (X_t - \hat{X}_t(\phi, \theta)) / (\gamma_{t-1}(\phi, \theta))^{1/2}; t = 1, 2, ..., n.$$

with $E(W_t(\phi, \theta) - z_t)^2 \to 0$ as $t \to \infty$.

Specifically, $\{\hat{W}_t\}$ is approximately

    (i) uncorrelated if $\{z_t\} \sim WN(0, \sigma^2)$
    (ii) independent if $\{z_t\} \sim IID(0, \sigma^2)$ and
    (iii) normally distributed if $z_t \sim N(0, \sigma^2)$

Dividing residuals $\hat{W}_t$ by $\hat{\sigma}$, we obtain rescaled residuals

$$\hat{R}_t = \hat{W}_t / \hat{\sigma}$$

which approximates either a WN(0,1) or an IID(0,1), provided $\{z_t\}$ of ARMA(p, q) process is independent WN.

These properties are easily verified and checked. If the selected ARMA(p, q) model is not the true model, then approximate modifications are made in model dimensions.

If the ARMA(p, q) selected is indeed the true model, and the residuals are iid, then residual autocovariants are iid with distribution $N(0, 1/n)$ for large n.

The 95% confidence boundaries are $\pm 1.96/\sqrt{n}$, which helps in our acceptance/rejection of ARMA(p, q) model.

The randomness of residuals can also be tested by fitting AR(p) model to residuals with $p = 0$ to 26 and finding the p value for which AICC is minimum.

If $p = 0$, then residuals are white noise.

A portmanteau χ^2 test for residual squared autocorrelation values may also be done assuming residuals are iid sequence for the large n.

2.3.3 Order Selection

Once the time series data is transformed to a stationary sequence $\{X_t\}$, we can fit a zero-mean ARMA model but we do not know exact order of p and q. The usual classical method of data transformation to a stationary sequence is made of combinations of Box-Cox normality transform and differencing the transforms or removal of trend and seasonal components. It is advantageous to choose p and q as small as possible so that the model is parsimonious ($p + q$ is minimum). This is convenient for forecasting as well. We should introduce a penalty factor to discourage models with too many parameters. Several criteria based on penalty factors arise in statistics and we concentrate on a few, such as FPE, AIC, BIC and AICC (bias-corrected AIC).

FPE Criterion

Akaike (1969) developed it for order selection of AR(p) model so that the one-step MSE for the model fitted to $\{X_t\}$ is used to predict an independent realization $\{Y_t\}$ of the same process which generated $\{X_t\}$. In such a case, $\hat{\phi}_1, ..., \hat{\phi}_p$ are obtained for $\{X_t\}$ and we derive $E(Y_{n-1} - \hat{\phi}_1 Y_n \cdots$

$- \hat{\phi}_p Y_{n+1-p}) \approx \sigma^2(1 + p/n)$ for independent realization $\{Y_t\}$.

If $\hat{\sigma}^2$ is ML of σ^2, then for large n, $n\hat{\sigma}^2/\sigma^2$ is approximately χ^2 with $(n - p)$ d.f.

Thus, $\text{FPE}_p = \hat{\sigma}^2 \dfrac{n+p}{n-p}$ (for $p = 0, 1, ..., 10$) for estimated mean square

prediction error of Y_{n+1}.

So, p which minimizes FPE_p, gives the order of AR(p) model.

AICC Criterion

Akaike (1973) developed a better criterion based on information (Kullback-Leibler measure) and is known as AIC. Later bias-corrected version of AICC (Hurvich and Tsai, 1989) was used, which is termed AICC and used here.

If X is an n-dimensional vector with pdf belonging to family $\{f(.; \psi)$, $\psi \in \overline{\psi} \}$, the Kullback-Leibler discrepancy but $f(.:, \psi)$ and $f(.; \theta)$ is given by

$$d(\psi|\theta) = \Delta(\psi|\theta) - \Delta(\theta|\theta)$$

where $\Delta(\psi|\theta) = E_\theta(-2 \ln f(X; \psi)) = \displaystyle\int_{R^n} -2 \ln(f(x; \psi) f(x; \theta) dx.$

In general, $d(\psi, \theta) \geq -2 \ln\left(\displaystyle\int_{R^n} \dfrac{f(x; \psi)}{f(x; \theta)} f(x; \theta) dx\right) = 0$ iff $f(x; \psi) = f(x; \theta)$.

Suppose $X_1, ..., X_n$ is for a Gaussian ARMA(p, q) with parameters $\theta = (\beta, \sigma^2)$.

Then, for large sample sizes $E \beta, \sigma^2\left(\dfrac{S_Y(\hat{\beta})}{\hat{\sigma}^2}\right) \approx \dfrac{2(p+q+1)n}{n-p-q-2}$

where Y is an independent realization of the true process with parameter θ.

Therefore, $-2 \ln L_X(\hat{\beta}, \hat{\sigma}^2) + 2 (p+q+1)n/(n-p-q-2)$ is an unbiased estimator of K-L index $E_\theta(\Delta(\hat{\theta}|\theta))$. We minimize AICC $(\hat{\beta})$ where

$$\text{AICC} (\beta) = -2 \ln L_X(\beta, X_X(\beta)/n) + 2(p+q+1)n/(n-p-q-2).$$

The AIC statistic is given by

$$\text{AIC} (\beta) = -2 \ln L_X(\beta, \delta_X(\beta)/n) + 2(p+q+1).$$

AICC and AIC are asymptotically equivalent as $n \to \infty$.

The AICC has a greater penalty for overfitting and is far better than AIC criterion.

Another criterion BIC for a zero-mean causal ARMA(p, q) is given by Akaike (1978).

$$\text{BIC} = (n - p - q)\,\ln[n\hat{\sigma}^2/(n - p - q)] + n(1 + \ln\sqrt{2\pi})$$

$$+ (p + q)\,\ln\left[\left(\sum\nolimits_{t=1}^{n} X_t^2 - n\hat{\sigma}^2\right)\Big/(p + q)\right]$$

where $\hat{\sigma}^2$ is MLE of white noise variance. BIC is a consistent estimator, i.e. $\hat{p} \to p$ and $\hat{q} \to q$ with probability 1 as $n \to \infty$ which is not valid for AICC or AIC.

However, AICC, AIC or FPE are asymptotically efficient for AR processes, which is not true for BIC. Efficiency is a desirable property as forecast errors would be small and confidence intervals would be narrow.

However, we do not know a priori p and q and many completing models can be useful for modelling. In such cases and for multivariate time series, AICC criterion provides us with a rational criterion as these are unbiased and have large penalties for overfitting. Additional points for order selection among competitive models are whether the residuals are white and whether the model is simple (easy to interpret in the context of the scientific discipline of the data). For example, if all except $m(\leq p + q)$ of the coefficients are constrained to be zeros, then AICC (β) can be redefined as

$$\text{AICC }(\beta) = -2 \ln L_X(\beta, S_X(\beta)/n) + 2(m + 1)n/(n - m - 2).$$

2.4 FORECASTING

Let $\{x_i\}$ be a stationary and invertible time series from an ARMA(p, q) process and we denote $\hat{x}_i(l)$ to be the forecast of x_{i+l} made at pivot time i. Then, $\hat{x}_i(l), l = 1, 2\ldots$ is called forecast errors of the model parameters will be very small and these may be assumed known and fixed for all future values of the series. We can then obtain the minimum mean square error forecasts as follows:

Forecast at lead l can be linearly predicted based on the values of residuals observed at current time i and previous times and is given by

$$\hat{x}_i(l) = \psi_l^* z_i + \psi_{l+1}^* z_{i-1} + \cdots$$

where the moving average ψ^* weights are to be estimated.

The MA form of ARMA(p, q) is given by

$$x_{i+l} = (z_{i+l} + \psi_1 z_{i+l-1} + \cdots + \psi_{l-1} z_{i+1}) + (\psi_l z_i + \cdots)$$

So, $E[\{x_{i+l} - \hat{x}_i(l)\}^2] = (1 + \psi_1^2 + \cdots + \psi_{l-1}^2)\sigma^2 + \sum_{j=l}^{\infty} (\psi_j - \psi_j^*)^2 \sigma^2$

which is obviously minimized when $\psi_j^* = \psi_j,$, $j = l, l+1, \ldots$

Therefore, $\hat{x}(l) = \psi_l z_i + \psi_{l+1} z_{i-1} + \cdots$ and forecast error at i for lead l is

$$e_i(l) = x_{j+l} - \hat{x}_{i+l} = z_{i+1} + \psi_1 z_{i+l-1} + \cdots + \psi_{l-1} z_{i+1}$$

Denoting E_i as conditional expectation at time i, we get

$$E_i[z_{i+l}] = \begin{cases} 0, & l > 0 \\ x_{i+l}, & l \le 0 \end{cases}$$

Therefore, we get $E_i[\hat{z}_i(l)] = \hat{z}_i(l)$, $l > 0$ and $E_i[e_i(l)] = 0$ for $l > 0$, so forecast errors have zero expectation value.

$$E_i[x_{i+l}] = \hat{x}_i(l), l > 0.$$

So, forecasts are unbiased. But, if the forecast function is off to one side, it remains there for short run and, therefore, forecast errors are correlated.

Example 21

Prove that forecast errors are correlated.

We have $e_i(l) = \sum_{j=0}^{l-1} \psi_j z_{i+l-j}$ and for $m > 0$ we have $e_i(l + m) = \sum_{j=0}^{l+m-1} \psi_j z_{i+l+m-j}$ with both the expected values (means) equal to zero.

So, $\mathrm{Cov}[e_i(l), e_i(l + m)] = \sum_0^{l-1} \psi_j \psi_{j+m} \sigma^2$

which is not zero and we would have correlation depending on l and m.

The one-step ahead of forecasts is white noises which are uncorrelated and these generate the process. But for various pivots, the forecasts for a given lead time l will often be correlated as seen below.

Example 22

We have $e_i(l) = \sum_{j=0}^{l-1} \psi_j z_{i+l-j}$

$$e_{i+m}(l) = \sum_{j=0}^{l-1} \psi_j z_{i+m+l-j}$$

So, if, $|m| \ge l$, $\mathrm{Cov}[e_i(l), e_i(l + m)] = 0$

But if, $|m| < l$, $\text{Cov}[e_i(l), e_i(l+m)] = \sum_{j=0}^{l-1-|m|} \psi_j \psi_{j+|m|} \sigma^2$

which is not zero in general.

The variance of forecast error at lead l is given by

$$V(l) = Var[e_i(l)] = (1 + \psi_1^2 + \cdots + \psi_{l-1}^2)\sigma^2$$

which is finite for all stationary processes as it is less than σ_x^2 and its $(100-\alpha)\%$ confidence intervals are obtained as

$$\hat{x}_i(l) \pm k_\alpha \left\{ \left(\sum_{j=0}^{l-1} \psi_j^2 \right) \sigma^2 \right\}^{1/2}$$

where k_α is the $(100-\alpha/2)\%$ part of the standard normal distribution.

Since forecasts are unbiased and have minimum MSE, they have minimum variance in the linear class of forecast functions.

Updating Forecasts

Taking conditional expectations at time i, for $l > 0$, the optimal forecast function is given by

$$\hat{x}_i(l) = \phi_1 E_i[x_{i+l-1}] + \cdots + \phi_p E_i[x_{i+l-p}] + 0 + \theta_1 E_i[z_{i+l-1}] + \cdots + \theta_q E_i[a_{i+l-q}]$$

Then, for $1 < l < p$, (q say),

$$\hat{x}_i(l) = \phi_1 \hat{x}_i(l-1) + \cdots + \phi_{l-1}\hat{x}_i(1) + \phi_l x_i + \cdots + \phi_p x_{i+l-p} + \theta_l z_i + \cdots + \theta_q z_{i+l-q}.$$

Therefore, if we have observations $x_1, \ldots, x_N$ for when forecasts up to lead L are required, we calculate $\hat{x}_N(l)$, $l = 1, \ldots, L$, recursive as shown in first row of Table 2.7 and if necessary a few z's can be set to zero for this purpose (Line III).

When x_{N+1} becomes available (i.e. at $t = N$), we have

$$z_{N+1} = x_{N+1} - \hat{x}_{N+1}$$

and then we can easily update forecasts for leads 1 to $L-1$ by equating

$$\hat{x}_{N+1}(l) = \hat{x}_N(l+1) + \psi_l z_{N+1}$$

But we do not have $\hat{x}_N(L+1)$ but by reapproximating the conditional forecast function again, we can get value of $\hat{x}_N(L+1)$ (Line II) and so on.

The 95% confidence limits for forecasts at leads 1 to L can be easily computed as λ_l, $l = 1, 2, \ldots, L$ as $\lambda_l = k_\alpha \left\{ \left(\sum_0^{l-1} \psi_j^2 \right) \sigma^2 \right\}^{1/2}$, $l = 1, 2, \ldots, L$

(Line III).

Table 2.7   Updating the forecasts from information up to time *i*

			Lead *l*					
i	x_i	z_i	1	2	....	*L*−1	*L*	
N	x_N	-	$\hat{x}_N(1)$	$\hat{x}_N(2)$		$\hat{x}_N(L-1)$	$\hat{x}_N(L)$	Line I
N+1	x_{N+1}	z_{N+1}	$\hat{x}_{N+1}(1)$	$\hat{x}_{N+1}(2)$		$\hat{x}_{N+1}(L-1)$	$\hat{x}_{N+1}(L)$	Line II
(100 − α)% confidence			λ_1	λ_2		λ_{L-1}	λ_L	Line III
limits ±								

$$\lambda_e = \pm k_\alpha \left\{ \left(\sum_0^{l-1} \psi_j^2 \right) \sigma^2 \right\}^{1/2}, l = 1, 2, ..., L$$

We can also determine the shape of forecast functions.

For time *i* + *l*, the ARMA(*p*, *q*) model gives

$$\phi_p(B)x_{i+l} = \theta_q(B)z_{i+l}$$

and taking conditional expectations for *l* > *q*, we have

$$\phi_p(B)\hat{x}_i(l) = 0$$

which can be solved with *p* pivotal (starting) values $\hat{x}_i(q - p + 1), ..., \hat{x}_i(q)$, where $\hat{x}_i(j) = x_{i+j}$, for $j \le 0$.

So at leads greater than (*q* − *p*), the $\hat{x}_i$ follows a function called the eventual forecast function (e.f.f.). The e.f.f. is easily obtained using pivotal values which are related to recent values of the process through the MA operator (see Fig. 2.10).

Example 23

Find the form of e.f.f. for an ARMA(1,2) process.

For *l* >2, $(1 - \phi B)\hat{x}_i(l) = 0$.

So, $\hat{x}_i(l) = A\phi^l$

where *A* is obtained for pivotal value $\hat{x}_i(2) = A\phi^2$.

So, e.f.f. is fully determined by the forecast at lead 2, and it decays geometrically.

$$\hat{x}_i(2) = \phi\hat{x}_i(1) + \theta_2 z_i$$

and $$\hat{x}_i(1) = \phi x_i + \theta_1 z_i + \theta_2 z_{i-1}$$

Suppose, $\phi = 1/2$ and $\theta_1 = -.8$; $\theta_2 = .4$; $x_i = 3.24$; $z_i = 0.64$ and $z_{i-1} = 0.95$.

Then $\quad \hat{x}_i(1) = 1.62 - .512 + .38 = 1.488$

$\qquad \hat{x}_i(2) = .744 + .256 = 1.000$

So, $\qquad A = 4$ and e.f.f. is shown in Fig. 2.10

ARMA $(1,2)$ with $\phi = 1/2$ and $\theta_1 = -.8; \theta_2 = .4$.

Box and Jenkins forecasts are very good if $N > 50$ as parameters are estimated without much error. This insensibility of forecasts with respect to estimation of error is very fortunate in practical applications and often Box-Jenkins forecasts are among the best forecasts.

Forecasting ARMA(p, q) Processes

For causal ARMA(p, q) processes $\phi_p(B)X_t = \theta_q(B)z_t$

where z_t is WN$(0, \sigma^2)$, the forecasting can be simplified by applying scaling by σ^{-1} as

$\qquad W_t = \sigma^{-1} X_t, t = 1, 2...., m.$

$\qquad W_t = \sigma^{-1} \phi(B)X_t, t > m.$

where $m = \text{Max}(p, q)$.

We also use $\theta_0 = 1$ and $\theta_j = 0$ for all $j > q$ and also $p, q \geq 1$.

The autocovariances $k(i, j) = E(W_i, W_j); i, j \geq 1$ are

$$k(i, j) = \begin{cases} \sigma^{-2}\gamma_X(i-j) & i \leq i, j \leq m \\ \sigma^{-2}\left[\gamma_X(i-j) - \sum_{r=1}^{p} \phi_x\gamma_X(r-|i-j|)\right] & \min(i,j) \leq m < \max(i,j) \\ \sum \theta_r\theta_r + |i-j| & \min(i,j) > m \\ 0 & \text{otherwise} \end{cases}$$

We also have

$$\begin{cases} \hat{W}_{n+1} = \sum_{j=1}^{N} \theta_{nj}(W_{n+1-j} - \hat{W}_{n+1-j}) & 1 \leq n < m \\ \hat{W}_{n+1} = \sum_{j=1}^{q} \theta_{nj}(W_{n+1-j} - \hat{W}_{n+1-j}) & n \geq m \end{cases}$$

where θ_{nj} and MSE $r_n = (W_{n+1} - \hat{W}_{n+1})^2$ are found by innovations algorithm, recursively.

The algorithm is

$$k\binom{11}{2} = v_0$$

$$\theta_{n,n-k} = v_k^{-1}\left(k(n+1, k+1) - \sum_{j=0}^{k-1} \theta_{k,k-j}\theta_{n,n-j} v_j\right); 0 \le k < n$$

and
$$v_n = k(n+1, n+1) - \sum_{j=0}^{n-1} \theta_{n,n-j}^2 v_j.$$

The θ_{nj} values are zero when both $n \ge m$ and $j > q$ because of innovations algorithm and since $k(r, s) = 0$ if $r > m$ and $|r - s| > q$ the linearity of predictor and equivalence of W_n and X_n processes suggest

$$X_t - \hat{X}_t = \sigma\left[W_t - \hat{W}_t\right] \text{ for } t \ge 1.$$

So, $\hat{W}_{n+1} = P_n W_{n+1}$ and $\hat{X}_{n+1} = P_n X_{n+1}$, P_n being linear predictor. Therefore,

$$\hat{X}_{n+1} = \begin{cases} \sum_{j=1}^{n} \theta_{nj}(X_{n+1-j} - \hat{X}_{n+1-j}), 1 \le n < m \\ \phi_1 X_n + \cdots + \phi_p X_{n+1-p} + \sum_{j=1}^{q} \theta_{nj}(X_{n+1-j} - \hat{X}_{n+1-j}), n \ge m \end{cases}$$

and $E(X_{n+1-j} - \hat{X}_{n+1-j}) = \sigma^2 E(W_{n+1} - \hat{W}_{n+1})^2 = \sigma^2 \gamma_n$

Prediction of AR(p)

$$\hat{X}_{n+1} = \phi_1 X_n + \cdots + \phi_p X_{n+1-p}; n \ge p$$

Prediction of MA(q)

Apply prediction to ARMA(p, q) with $\phi_1 = 0$.

$$\hat{X}_{n+1} = \sum_{j=1}^{\min(n,q)} \theta_{nj}(X_{n+1-j} - \hat{X}_{n+1-j}), n \ge 1.$$

Here $k(i, j) = \sigma^{-2}\gamma_X(i - j) = \sum_{r=0}^{q-|i-j|} \theta_r \theta_{r+|i-j|}.$

Prediction of ARMA(1,1)

$$\gamma_X(0) = \sigma^2(1 + 2\theta\phi + \theta^2)/(1 - \phi^2)$$

We have

$$k(i,j) \;=\; \begin{cases} (1+2\theta\phi+\theta^2)/(1-\phi^2); & i=j=1 \\ 1+\theta^2; & i,j\geq 2 \\ \theta; & |i-j|=1, i\geq 1 \\ 0; & \text{otherwise} \end{cases}$$

Then, we obtain as final result

$$\gamma_0 \;=\; (1+2\theta\phi+\theta^2)/(1-\phi^2)$$

$$\theta_{n1} \;=\; \theta/\gamma_{n-1}$$

$$\gamma_n \;=\; (1+\theta^2-\theta^3)/\gamma_{n-1}$$

2.5 PURELY SEASONAL ARMA MODELS ARMA(P, Q)s

If it is a purely seasonal ARMA process, the dependence on the past tends to occur most strongly at multiples of some seasonal lags.

For example, normal data would have $s = 12$, quarterly data $s = 4$, weekly data $s = 7$ but natural phenomena such as temperature has a strong dependence on seasons but exact values of s are not fixed but may vary from place to place. So, we have purely seasonal ARMA(P, Q)s as:

$$\Phi_P(B^s)x_t \;=\; \theta_Q(B)z_t$$

where $\quad \Phi_P(B^s) \;=\; 1-\Phi_1(B^s)-\Phi_2(B^{2s})-\cdots-\Phi_P(B^{Ps})$

and $\quad \theta_Q(B^s) \;=\; 1+\theta_1 B^s+\theta_2 B^{2s}+\cdots+\theta_Q B^{Qs}.$

This is causal only when roots of $\Phi_P(B^s)$ lie outside the unit circle and it is invertible only if the roots of $\theta_Q(B^s)$ lie outside the unit circle.

For example, we consider ARMA$(1,1)_{12}$ model

$$(1-\Phi B^{12})x_t \;=\; (1+\theta B^{12})z_t$$

or $\qquad x_t \;=\; \Phi x_{t-12}+z_t+\theta z_{t-12}.$

It is clear that the causal condition requires $|\Phi| < 1$ and invertible condition requires $|\theta| < 1$.

The MA$(1)_{12}$ model will have

$$\gamma(0) \;=\; (1+\theta^2)\sigma^2$$

$$\gamma(\pm 12) \;=\; \theta\sigma^2$$

$$\gamma(h) \;=\; 0 \text{ otherwise.}$$

Thus, the only non-zero correlation except at lag 0, is given by

$$\rho(\pm|z|) \;=\; \Theta/(1+\Theta^2).$$

PACF can be calculated.

Similarly, AR(1)$_{12}$ model will have

$$\gamma(0) = \sigma^2/(1 + \Phi^2)$$

$$\gamma(\pm 12k) = \sigma^2\Phi^k/(1 - \Phi^2); k = 1, 2, ...$$

$$\gamma(h) = 0 \text{ otherwise.}$$

So, the only non-zero correlations are

$$\rho(\pm 12k) = \Phi^k; k = 0, 1, 2, ... \text{ (PACF can be calculated).}$$

The initial diagnostics of purely seasonal ARMA(P, Q)s are given in the Table 2.8.

Table 2.8

	AR(P)s	MA(Q)s	ARMA(P, Q)
ACF#	Tails off at lag ks $k = 1, 2,...$	Cut off after lag Qs	Tails off at lag ks
PACF#	Cuts off after lag Ps	Tails off at lag ks $k = 1, 2,...$	Tails off at lag ks

Note: Values are zero at lags $h \neq k$s for $k = 1, 2, 3 ...$

Consider a mixed seasonal model ARMA(0,1) × (1,0)$_{12}$, i.e.

$$x_t = \Phi x_{t-12} + z_t + \theta z_{t-1}$$

where $|\Phi| < 1$ and $|\theta| < 1$.

We get $\gamma(0) = \Phi^2\gamma(0) + \sigma^2 + \theta^2\sigma^2$ or $\gamma(0) = \left[(1 + \theta^2)/(1 - \Phi^2)\right]\sigma^2$

$$\gamma_1 = \Phi\gamma(11) + \theta\sigma^2$$

and $\gamma(h) = \Phi(h - 12), h \geq 2.$

Thus, ACF is given by

$$\rho(12h) = \Phi^h; h = 1, 2,...$$

$$\rho(12h - 1) = \rho(12h + 1) = \left(\frac{\theta}{1 + \theta^2}\right)\Phi^2; h = 0, 1, 2...$$

$$\rho(h) = 0, \text{ otherwise.}$$

The ACF and PACF for typical values $\Phi = 0.8$ and $\theta = -.5$ are shown in Table 2.9:

Example 24   *(Determining d and D in SARIMA models)*

Suppose the variate differencing yielded the following matrices for different time series data:

(i) Airline passenger, International lines

(ii) Women unemployment

(iii) Passenger miles flown

(iv) Log sales data

(v) Mean monthly air temperature

Table 2.9

(i) $D_{12} =$ 0   1   2

$$\begin{array}{c} d=0 \\ =1 \\ =2 \end{array} \begin{pmatrix} .1949 & .0038 & .0100 \\ .0114 & \mathbf{.0021} & .0060 \\ .0182 & .0057 & .0157 \end{pmatrix}$$

0   1   2

(ii) $\begin{pmatrix} 254 & 244 & 108 \\ 23 & \mathbf{8} & 17 \\ 32 & 11 & 19 \end{pmatrix}$

(iii) $\begin{pmatrix} 1076 & 44 & 78 \\ 332 & \mathbf{34} & 91 \\ 250 & 87 & 226 \end{pmatrix}$

(iv) $\begin{pmatrix} .520 & \mathbf{.054} & .142 \\ .107 & \mathbf{.061} & .163 \\ .139 & .188 & .520 \end{pmatrix}$

(v) $\begin{pmatrix} 73 & \mathbf{12} & 39 \\ 27 & 19 & 65 \\ 30 & 55 & 190 \end{pmatrix}$

Answer:

(i) series $d = 1$, $D_{12} = 1$,   (ii) series $d = 1$, $D_{12} = 1$

(iii) series $d = 1$, $D_{12} = 1$,   (iv) series $d = 0$, $D_{12} = 1$

(v) series $d = 0$, $D_{12} = 1$.

Example 25

Obtain theoretical acf for $x_i = (1 + \theta B)(1 + \Theta B^{12})z_i$.

Give the invertibility region.

Ans: $\rho_1 = \theta/(1 + \theta^2)$ for ARIMA part.

$\rho_{12} = \Theta/(1 + \Theta^2)$ for SARIMA part.

$\rho_{11} = \theta\Theta/(1 + \theta^2)(1 + \Theta^2) = \rho_{13}$ (corrected by intervention term ($= \rho_1\rho_{12}$)).

Otherwise for $k > 1$, $\rho_k = 0$.

The process is invertible if $-1 < \theta, \Theta < 1$ when the zeros of the multiplicative MA all lie outside the unit circle. The acf for multiplicative SARIMA is easily obtained by autocovariance generating function (AGF)

$$F(x) = \sum_{-\infty}^{\infty} \gamma_k x^k$$

where $\{\gamma_k\}$ is the appropriate autocovariance function.

If the AGF of two processes Γ_1 are and Γ_2, the AGF of multiplicative process is $\Gamma_1\Gamma_2$ because for any linear model $x_i = \psi(B)z_i$ has $\Gamma(x) = \sigma^2\psi(t)\psi(t^{-1})$.

Effective length n of the series for a multiplicative SARIMA is

$$n = N - d - DS.$$

A more general model is obtained by replacing $\theta\Theta$ with θ^* in the above example, so the model for x_i is

$$x_i = z_i + \theta z_{i-1} + \Theta z_{i-12} + \theta^*_{i-13}$$

However, if estimation shows that $\theta\Theta - \theta^*$ is not significantly different from zero, this modification is unnecessary.

The threshold autocovariances for this general model are

$$\gamma_0 = 1 + \theta^2 + \Theta^2 + \theta^{*2}, \qquad \gamma_1 = \theta + \Theta\theta^*$$

$$\gamma_{12-1} = \theta\Theta, \qquad \gamma_{12} = \theta\theta^* + \Theta$$

$$\gamma_{12+1} = \theta^* \text{ and otherwise for } k > 1, \gamma_k = 0.$$

If such non-multiplicative model is required, it is best obtained as a suitable modification of the closest fitting multiplicative model, which is later shown to be inadequate at the verification stage. Then the non-multiplicative seasonal model would be required and considered to be adequate over the discarded multiplicative model.

2.6 MODEL GENESIS AND REALIZATION

In many sciences, often incompletely understood theoretical consider-ations are used to decide on a preferred model over the statistical results from the data. However, statistical models such as the Box-Jenkins ones often yield smaller mean-square errors compared to threshold models as well as these do not require data on additional external variables for modelling and/or forecasting. Often, in earth sciences we observe results of the sum of several independent processes occurring over a period of time and/or space. Hence, the problem would be to decompose observed models into their original independent processes,

i.e. if $\{z_i\} \sim ARMA(p, q)$ and $z_i = x_i + y_i$, where $\{x_i\}$ is $ARMA(p_1, q_1)$ and $\{y_i\}$ is $ARMA(p_2, q_2)$ in the stationary case.

In the non-stationary case $z_i \sim ARIMA(p, d, q)$ and $x_i \sim ARIMA(p_1, d_1, q_1)$ and $y_i \sim ARIMA(p_2, d_2, q_2)$.

It is easy to solve forward modelling when x_i and y_i are known models and model for z_i is possible to compute using Granger's (1972) lemma:

$$MA(q_1) + MA(q_2) = MA(q)$$

where $q \le \max[q_1, q_2] = q^*$ (say).

Then, it is easy to prove the main theorem for superposed models (Anderson, 1976, p. 136):

$$ARMA(p_1, q_1) + ARMA(p_2, q_2) = ARMA(p, q)$$

where $p = p_1^* + p_2^* + h$ and $q \le \max[p_2^* + q_1, p_1^* + q_2]$ and h being the degree $h(B)$ polynomial which is highest common factor for the two AR operators,

i.e. $\phi_{p_1}(B) = h(B)\,\phi_{p_1^*}(B)$ and $\phi_{p_2}(B) = h(B)\,\phi_{p_2^*}(B)$.

For interpreted models, this theorem for $d_1 \ge d_2$ becomes

$$ARIMA(p_1, d_1, q_1) + ARMA(p_2, d_2, q_2) = ARIMA(p, d_1, q)$$

where $p = p_1^* + p_2^* + h$ and $q \le \max[p_2^* + q_1, p_1^* + d_1 - d_2 + q_2]$.

Superposed models can be similarly solved.

The main theorems for superposed models show the reason why $ARMA(p, q)$ or $ARIMA(p, d, q)$ models are found more often than pure AR or pure MA processes. For example, restricting to $(p + q) \le 2$, we obtain the following possible models [models below the diagonal are omitted because of symmetry].

Table 2.10

$x \rightarrow$ $y \downarrow$	WN	MA(1)	MA(2)	AR(1)	AR(2)	ARMA(1,1)
WN	WN	MA(1)	MA(2)	ARMA(1,1)	ARMA(2,2)	ARMA(1,≤1)
MA(1)		MA(≤1)	MA(2)	ARMA(1,2)	ARMA(2,3)	ARMA(1,≤2)

(Table 2.10 Contd.)

(Table 2.10 Contd.)

	MA(≤2)	ARMA(1,3)	ARMA(2,4)	ARMA(1,≤3)
MA(2)	MA(≤2)	ARMA(1,3)	ARMA(2,4)	ARMA(1,≤3)
AR(1)		ARMA(2,≤1)	ARMA(2,≤1 Or 3,≤2)	ARMA(1,≤1 Or 2,≤2)
AR(2)			ARMA(3,≤1, Or 4,≤2)	ARMA(2,≤2 Or 3,≤3)
ARMA(1,1)				ARMA(1,≤1 Or 2,≤2)

We should be careful to question whether a particular ARMA(p, q) model is really a sum of heterogeneous quantities each following simpler, distinct and independent processes; or is the process simple but contaminated by a observational error series. In realizing tectonic processes, one is faced with the addition of several independent tectonic phases where each tectonic phase may have simpler models but the superposition of these models yield more complex ARMA(p, q) models. We should also examine if these simple situations are realizable. For example, can a ARMA(1,1) model be sum of AR(1) and white noise? If so, $0 < -\{\theta(1 + \phi^2)/\phi(1 + \theta^2)\} < 1$.

This is equivalent to $- \theta\phi > 0$ and < 1, we obtain the realizability condition as

$$|\theta| < |\phi|$$

Superposed ARMA(p, q) Models

In geological sciences, there are many superposed phenomena with respect to time/spatial dimension and both in time and space. A few examples could be superposition by diagenetic, mineralization, metamorphic and hydrothermal mixing processes, where we observe the complex superposed effects of the original simple processes, assuming original processes are stationary and invertible ARMA(p, q). However, we have to find the relation between p and q in terms of originals p_1, q_1 and p_2, q_2, i.e. if $\{x_i\}$ and $\{y_i\}$, the superposed process is

$$\{z_i\} = \{x_i\} + \{y_i\}$$

and   ARMA(p, q) = ARMA(p_1, q_1) + ARMA(p_2, q_2).

Before we can relate p to p_1, q_1, p_2, q_2; we consider the superposition of simpler MA(q) processes, where $\rho_k = 0$ for $k > q$ which is not a solution condition. Wold (1954) has given the necessary and sufficient conditions for $\rho_k = 0$, $k > q$ as it is positive if $\rho(\zeta)$ has no roots of odd multiplicity in open range (−1,1) where

$$\rho(\cos w) = 1 + 2\sum\nolimits_{j=1}^{q} \rho_j \cos jw$$

So, $\rho(\zeta) \geq 0$ for all $\zeta \in (-1,1)$.

To be invertible $\rho(1) = 1 + 2\sum\nolimits_{j=1}^{q} \rho_j$ must be positive and greater than one (by putting $w = 0$ in equation $\rho(\cdot)$).

But according to Wold, there should not be root $\rho(1)$ within $(-1,1)$, which is always true.

Therefore, we can prove Granger's (1972) Lemma:

$$MA(q_1) + MA(q_2) = MA(q)$$

where $q \leq \max[q_1, q_2] = q_*$(say).

If the MA processes on the left side are invertible, we have

$$\sigma^2\rho(\zeta) = \sigma_1^2\rho_1(\zeta) + \sigma_2^2\rho_2(\zeta)$$

are non-negative over $(-1,+1)$.

So, $\rho(\zeta)$ does not cross the axis in $(-1,1)$ and does not have root of odd multiplicity there.

Hence, $MA(q)$ is also invertible.

Therefore, if $\{x_i\}$ and $\{y_i\}$ are independent processes

$$\text{Cov}[z_i, z_{i-k}] = \text{Cov}[x_i, x_{i-k}] + \text{Cov}[y_i, y_{i-k}]$$

Hence, for $MA(q_1) + MA(q_2)$, $\rho_k = 0$ for $k > q_*$.

Even if $\{x_i\}$ and $\{y_i\}$ are not independent processes, it is still valid that (linear) sum of $MA(q_1)$ and $MA(q_2)$ is invertible $MA(q \leq q^*)$.

If the sum of two processes is denoted by w_i with autocovariance $\gamma_k(w)$ and parameters $\theta_j(w)$, we have for $k = q_*$.

$$\gamma_{q_*}(z) = \gamma_{q_*}(z) + \gamma_{q_*}(y) = \theta_{q_*}(x)\,\sigma^2(x) + \theta_{q_*}(y)\,\sigma^2(y)$$

So, for $q < q_*$; if (i) $q_1 = q_2 = q_*$ and (ii) $q_q(x)\,\sigma^2(x) + \theta_{q_*}(y)\sigma^2(y) = 0$, we have a model for w as $MA(q_*) = MA(q)$.

Using above lemma, we can prove the main theorem.

$$ARMA(p_1, q_1) + ARMA(p_2, q_2) = ARMA(p, q).$$

where $p = p_1^* + p_2^* + h$ and $q \leq \max[p_2^* + q_1, p_1^* + d_1 - d_2 + q_2]$; $h(B)$ is the polynomial in B and is highest common factor of the two AR polynomials $\phi_{p1}(B)$, $\phi_{p2}(B)$.

Let $\phi_{p1}(B)x_i = \theta_{q1}(B)a_i$ and $\phi_{p2}(B)y_i = \theta_{q2}(B)b_t$,

where $\{a_i\}$, $\{b_i\}$ are independent white noise processes and $z_i = x_i + y_i$

Therefore,

$$z_i = \phi_{p_1}^{-1}(B)\,\theta_{q_1}(B)a_i + \phi_{p_2}^{-1}(B)\,\theta_{q_2}(B)b_i$$

$$= h^{-1}(B)[\,\phi_{p_1}^{-1}(B)\,\theta_{q_1}(B)a_i + \phi_{p_2}^{-1}(B)\,\theta_{q_2}(B)b_i]$$

So, $h(B)\,\phi_{p_1}^{-1}(B)\,\phi_{p_2}^{-1}(B)z_i = \phi_{p_2}^{-1}(B)\,\theta_{q_1}(B)a_i + \phi_{p_1}^{-1}(B)\,\theta_{q_2}(B)b_i$.

On the RHS, we have $MA(p_2^* + q_1) + MA(p_1^* + q_2)$

which is summed to yield $MA(q \le \max[p_2^* + q_1, p_1^* + q_2]$.

The AR operator in LHR is of degree $h + p_1^* + p_2^*$.

So, we have $z_i \sim ARMA(p_2^* + p_1^* + h, \le \max\,[p_2^* + q_1, p_1^* + q_2]\,)$ with

strict equations only if $p_2^* + q_1 = p_1^* + q_2$.

(*cf*: Table 2.11 for $p + q$ unrestricted and $p + q \le 2$.)

Table 2.11   Sum of independent ARMA processes, 1+2. Models below diagonal are not shown but easily obtained

$1 \rightarrow$ $2 \downarrow$	WN	$MA(q_1)$	$AR(p_1)$	$ARMA(p_1, q_1)$
WN	WN	$MA(q_1)$	$ARMA(p_1, q_1)$	$ARMA(\,)$ $(p_1, \le \max[q_1, p_1]$
$MA(q_2)$		MA	ARMA	ARMA
		$(q \le \max[q_1, q_2]$	$(p_1, p_1 + q_2)$	$(p_1^* + p_2^* - h,$ $\le [q_1, p_1 + q_2])$
$AR(p_2)$			ARMA	ARMA
			$(p_1^* + p_2^* + h,$ $\le [p_2^*, p_1^*]$	$(p_1^* + p_2^* - h,$ $\le \max[p_2^* + q_1, p_1^*])$
ARMA				ARMA
(p_2, q_2)				$(p_1^* + p_2^* + h,$ $\le \max[p_2^* + q_1, p_1^* + q_2])$

Restricting to $p + q \le 2$ we can use the above superposed stationary models to give the data as shown in Table 2.12 (without giving the models below the diagonal which can be easily computed):

Table 2.12

$1 \rightarrow$ $2 \downarrow$	WN	MA(1)	MA(2)	AR(1)	AR(2)	ARMA(1,1)
WN	WN	MA(1)	MA(2)	ARMA(1,1)	ARMA(2,2)	ARMA(1, ≤ 1)
MA(1)		MA(1, ≤ 1)	MA(2)	ARMA(1,2)	ARMA(2,3)	ARMA(1, ≤ 2)
MA(2)			MA(1, ≤ 2)	ARMA(1,3)	ARMA(2,4)	ARMA(1, ≤ 3)
AR(1)				ARMA(2, ≤ 1)	ARMA(2, ≤ 1 Or 3, ≤ 2)	ARMA(1, ≤ 1 Or 2, ≤ 2)
AR(2)					ARMA(3, ≤ 1 Or 4, ≤ 2)	ARMA(2, ≤ 2 Or 3, ≤ 3)
ARMA (1,1)						ARMA(1, ≤ 1 Or 2, ≤ 2)

2.7 EXAMPLES

Example 26 *(Sunspot Numbers)*
The sunspot numbers for years 1770 to 1869 have mean 46.93 and autocovariances $\hat{\gamma}(0) = 1382.2$, $\hat{\gamma}(1) = 1114.4$, $\hat{\gamma}(2) = 591.73$, and $\rho_0 = 1$, $\rho_1 = .8$, $\rho_2 = .43$.

The PACF are $\phi_{11} = 0.8$, $\phi_{22} = -0.6$ and ϕ_{kk} nearly zero (non-significant) for all higher k (SE $\phi_{kk} = \pm 0.196$).

This shows that an AR(2) model is suitable and can be obtained by solving Yule-Walker equations

$$X_t = 1.318\, X_{t-1} - .634\, X_{t-2} + z_t;\ z_t \sim WN(0,289.2).$$

Example 27 *(Glacial Varve series)*
The thickness of the glacial varve units are couplets with (coarser and thicker) sand (and silt) strata deposited in summer and (finer and thinner) shales with organic material deposited in winter. They, therefore, form a time series in thickness distribution, which is lognormal. The log thickness values, however, exhibit linear trends which can be made stationary by differencing.

Thus, we have

$$\nabla \ln(x_t) = \ln(x_t/x_{t-1})$$

which is logarithm of proportional change in thickness as being stationary.

The acf of $\nabla \ln(x_t) = \ln(x_t/x_{t-1})$ is significantly negative at ρ_1 and non-significant for higher lags, whereas pacf shows that the exponential decreases with higher lags. Thus, an MA(1) model seems suitable with $\hat{\theta} = -.772$ and $\sigma^2 = 0.236$.

The model $\ln\left(\dfrac{x_1}{x_{t-1}}\right) = z_t - .772_{t-1}$

where $z_t \sim WN(0,0.236)$.

Example 28   *(Asymptotic discriminants)*

AR(1): $\gamma_x(0) = \sigma^2/(1 - \phi^2)$, so $\sigma^2\Gamma_{1,0}^{-1} = (1 - \phi^2)$

Thus, $\hat{\phi} \sim AN[\phi, n^{-1}(1 - \phi^2)]$.

AR(2): $\gamma_x(0) = \dfrac{(1 - \phi_2)\sigma^2}{(1 + \phi_2)(1 - \phi_2)^2 - \phi_1^2}$

and $\gamma_x(1) = \phi_1\gamma_x(0) + \phi_2\gamma_x(1)$.

We can compute $\Gamma_{2,0}^{-1}$.

$$\begin{pmatrix}\hat{\phi}_1 \\ \hat{\phi}_2\end{pmatrix} \sim AN\left[\begin{pmatrix}\phi_1 \\ \phi_2\end{pmatrix}, n^{-1}\begin{pmatrix}1 - \phi_2^2 & -\phi_1(1 + \phi_2) \\ -\phi_1(1 + \phi_2) & 1 - \phi_2^2\end{pmatrix}\right].$$

MA(1):   We have $\theta(B)y_t = z_t$.
Analogous to AR(1), we have $\gamma_y(0) = \sigma_z^2/(1 - \theta^2)$.

So, $\sigma^2\Gamma_{0,1}^{-1} = (1 - \theta^2)$.

Thus, $\hat{\theta} \sim AN[\theta, n^{-1}(1 - \theta^2)]$.

MA(2): $y_t + \theta_1 y_{t-1} + \theta_2 y_{t-2} = z_t$.
So, analogous to AR(2), we have

$$\begin{pmatrix}\hat{\theta}_1 \\ \hat{\theta}_2\end{pmatrix} \sim AN\left[\begin{pmatrix}\theta_1 \\ \theta_2\end{pmatrix}, n^{-1}\begin{pmatrix}1 - \theta_2^2 & -\theta_1(1 + \theta_2) \\ -\theta_1(1 + \theta_2) & 1 - \theta_2^2\end{pmatrix}\right].$$

ARMA(1,1):   To calculate $\Gamma_{\phi\theta}$, we must find $\gamma_{xy}(0)$, where $x_t - \phi x_{t-1} = z_t$
and $y_t - \theta y_{t-1} = z_t$.

We have $\gamma_{xy}(0) = Cov(x_t, y_t) = Cov(\phi x_{t-1} + z_t, -\theta y_{t-1} + z_t)$
$= -\phi\theta\gamma_{xy}(0) + \sigma^2$.

$$\gamma_{xy}(0) = \sigma^2/(1 + \phi\theta).$$

and $\begin{pmatrix}\hat{\phi} \\ \hat{\theta}\end{pmatrix} \sim AN\left[\begin{pmatrix}\phi \\ \theta\end{pmatrix}, n^{-1}\begin{pmatrix}(1 - \phi^2)^{-1} & (1 + \phi\theta)^{-1} \\ (1 + \phi\theta)^{-1} & (1 - \theta^2)^{-1}\end{pmatrix}^{-1}\right].$

The asymptotic results are very good for large samples but when sample size is small or when parameters are close to the boundaries of unit circle, the asymptotic approximations can be quite poor.

2.8 EXERCISES

1. Let $\{x_t\}$ be stationary and we apply two filters in succession as $y_t = \sum_r a_r x_{t-r}$. Then, $z_t = \sum_s b_s y_{t-s}$.

   Show the spectrum of output is $f_z(v) = |A(v)|^2 + |B(v)|^2 f_x(v)$, where $A(v)$ and $B(v)$ are Fourier transforms of filters a_t and b_t, respectively.

2. We apply filter $u_t = x_t - x_{t-1}$ to an economic time series x_t and follow it by filter $v_t = u_t - u_{t-12}$.

   What could be the model for v_t time series? Is u_t stationary?

   Ans: The first filter attenuate low-frequency trends while the second one attenuate frequencies at surpluses of monthly periods.

3. Consider two time series $x_t = z_t - z_{t-1}$ and $y_t = \frac{1}{2}(z_t + z_{t-1})$, $\{z_t\} \sim WN(0,1)$.

   (a) Are u_t and y_t jointly stationary? (Cross covariance function is function of log h and independent of t).

   (b) Compute spectra $f_y(v)$ and $f_x(v)$ and comment on the differences in their spectra.

4. Periodicities of sunspot series number can be judged by fitting an autoregressive spectrum of high order and generally assumed to be 11 years. Can you suggest any other competing model along with your reasons for the same?

5. If $y_t = \sum_{r=-\infty}^{\infty} a_r x_{t-r}$ (without any random errors), then show that square coherence $\rho_{y,x}^2(v) = 1$ for all v.

6. Let X, Y be two random variables with $E(Y) = \mu$ and $EY^2 < \infty$.

   If constant C minimizes $E(Y - C)^2$, then show $C = \mu$.

   If random variable $f(X)$ minimizes $E\sum (Y - f(X))^2$, then show that

   $$f(X) = E[Y \mid X].$$

7. Show that a strictly stationary process $E(X_i^2) < \infty$ is weakly stationary.

8. Let $\{X_t\}$ be MA(1) process with parameter θ and $\{z_t\}$ is WN(0,1). Compute ACVF and ACF when $\theta = 0.8$. Compute variance of sample mean $\overline{X} = \sum_{t=1}^{4} X_t / 4$ when $\theta = 0.8$ and when $\theta = -0.8$.
   Compare the last two results.

9. Let $\{X_t\}$ be AR(1) with $\phi = 0.9$ and $\sigma^2 = 1$. Compute variance and sample mean $\overline{X} = \sum_{t=1}^{4} X_t / 4$. Compute variance of $\overline{X}$ when $\phi = -.9$, $\sigma^2 = 1$ and compare these two results.

10. Find a polynomial filter that passes the linear trends without distortion and estimates seasonal components of period 2.

11. If $\{Y_t\}$ is a stationary process with mean zero and $X_t = a + bt + s_t + Y_t$, where a, b are constants and s_t is seasonality with period 12, prove that $\nabla\nabla_{12}X_t$ is stationary. Find its acvf in terms of that of $\{Y_t\}$. If $X_t = (a + bt) s_t + Y_t$, then show that $\nabla_{12}^2 X_t$ is stationary. Find its acvf in terms of that of $\{Y_t\}$.

12. Show that $\dfrac{1}{1-\phi z} = \sum_{j=1}^{\infty} \phi^{-j} z^{-j}$ for $|\phi| > 1$ and $|z| \geq 1$.

13. Let $\{Y_t\}$ be AR(1) plus noise time series defined as $\{Y_t\} = X_t + W_t$, where $\{W_t\} \sim WN(0, \sigma_N^2)$ and $X_t - \phi X_{t-1} = z_t$, $\{z_t\} \sim WN(0, \sigma_z^2)$ and $E(W_s z_t) = 0$ for all s and t.
   Show that $\{Y_t\}$ is stationary and compute its acvf.
   Show that $Y_t - \phi Y_t - 1$ is lag 1-correlated, i.e. MA(1) process. Show that $\{Y_t\}$ is ARMA(1,1) and compute its three parameters in terms of ϕ, σ_w^2, σ_z^2.

14. Let $\{X_t\}$ be a stationary process with mean and μ acf $\rho(\cdot)$
   Show that the best linear predictor of X_{n+b} is obtained when slope $a = \rho(h)$ and constant $b = \mu(1 - \rho(h))$.

15. Let $\{X_t\}$ be a stationary AR(p) process with $\{z_t\} \sim WN(0, \sigma^2)$ and uncorrelated with for X_s each $s < t$. Then the best linear predictor of X_{n+1} in terms of 1, $X_1, \ldots, X_n$ $(n > p)$ is given by $\rho_n X_{n+1} = \phi_1 X_n + \cdots + \phi_p X_{n+1-p}$. Find the MSE of $\rho_n X_{n+1}$.

16. Let $\{Y_t\}$ be ARMA(p, q) plus noise defined by $Y_t = X_t + W_t$, where $\{W_t\} \sim WN(0, \sigma_w^2)$ and $\phi_p(B)X_t = \theta_q(B)z_t$ and $\{z_t\} \sim WN(0, \sigma_z^2)$ with $E(W_s z_t) = 0$ for all s and t.

Show $\{Y_t\}$ is stationary. Find its acvf in terms of σ_w^2 and the acvf of $\{X_t\}$. Show that $\{Y_t\}$ is ARMA(p, max (p, q)).

17. Let $\{X_t\}$ be non-invertible MA(1) process $X_t = z_t + \theta z_{t+1}$ with $\{z_t\} \sim$ WN(0, σ^2) and $|\theta| > 1$. Show that $\{W_t\} \sim$ WN(0, σ_w^2), where $W_t =$

$$\sum_{j=0}^{\infty} (-\theta)^{-j} X_{t-j}.$$

Compute σ_w^2 in terms of θ and σ^2 and show that $\{X_t\}$ has invertible

representation $X_t = W_t + \dfrac{1}{\theta} W_{t-1}$.

18. Calculate the pacf of MA(1) process $X_t = z_t + \theta z_{t+1}$ at lag 2, where $\{z_t\} \sim$ WN(0, σ^2) and $t = 0, \pm 1, \pm 2, ...$
    (Ans: $-\theta^2/(1 + \theta^2 + \theta^4)$).

19. Compute the first six coefficients ψ_i, $i = 0, ..., 5$ in causal

representation of $X_t = \displaystyle\sum_{j=0}^{\infty} \psi_j z_{t-j}$ of $\{X_t\}$.

   (i) $X_t = -.2 X_{t-1} + .48 X_{t-2} + z_t$

   (ii) $X_t = -.6 X_{t-1} + z_t + 0.4 z_{t-1}$

   (iii) $X_t = -1.8 X_{t-1} - 0.81 X_{t-2} + z_t$

20. Find the spectral density function of white noise $\{z_t\} \sim$ WN(0, σ^2).

21. If $\{X_t\}$ and $\{Y_t\}$ are uncorrelated stationary processes and $\{z_t\} = \{X_t + Y_t\}$, show that the $\{z_t\}$ process is stationary with acvf $\gamma_z = \gamma_X + \gamma_Y$ and spectral distribution function $F_z = F_X + F_Y$.

22. If $X_t = A \cos (\pi t/3) + B(\pi t/3) + Y_t$, where $Y_t = z_t + 2.5 z_{t-1}$; $\{z_t\} \sim$ WN(0, σ^2), A, B are uncorrelated with mean 0 and variance v^2, $\{z_t\}$ is uncorrelated with A and B for each t, then find the acvf and spectral density function of $\{X_t\}$.

23. The spectral density function of a real-valued time series $\{X_t\}$ in $[0, \pi]$ is given by

$$f(\lambda) = \begin{cases} 100 & \text{if } (\pi/6 - .01 < \lambda < \pi/6 + .01) \\ 0 & \text{otherwise} \end{cases}$$

and on $[-\pi, 0]$ by $f(\lambda) = f(-\lambda)$.

Find the acvf of $\{X_t\}$ at lags 0 and 1. Find the variance and spectral density functions of $\{Y_t\}$, where $Y_t = X_t - X_{t-12}$. Find the transfer function of filter ∇_{12} and indicate its effects at frequencies of sinusoids at $\lambda = 0$ and $\lambda = \pi/6$.

24. Find the causal model of ϕ values for the following AR(2) process $X_t - \phi X_{t-1} - \phi^2 X_{t-2} = z_t$, $\{z_t\} \sim WN(0, \sigma^2)$.

If 200 data of one realization yields $\hat{\gamma}(0) = 6.06$ and $\hat{\rho}(1) = .687$, find estimates of ϕ and σ^2 correspondingly to the causal ϕ model.

25. We have two observations x_1 and x_2 from the causal AR(1) process $X_t = \phi X_{t-1} + z_t$, $\{z_t\} \sim WN(0, \sigma^2)$ and $|x_1| \neq |x_2|$.

Find the maximum likelihood estimator for ϕ and σ^2.

References

1. Akaike, H., 1973, Information theory and an extension of Maximum Likelihood Principle, 2nd *International Symposium Inf. Theory*, Akadmiaikiado, Budapest, pp. 267-281.

2. Anderson, T.W., 1971, *Statistical Analysis of Time Series*, Wiley, New York.

3. Bartlett, M.S., 1946, On the theoretical specification of sampling properties of autocorrelated Time Series, *Journal Royal Statistical Society*, B.8: 27-41.

4. Bloomfield, P., 1976, *Fourier Analysis of Time Series*, Wiley, New York.

5. Box, G.E.P. and Cox, D.R., 1964, An Analysis of Transformations, *Journal Royal Statistical Society*, B26: 211-252.

6. Dempster, A.P., Laird, N.M. and Rubin, D.B., 1977, Maximum likelihood for incomplete data via EM algorithm, *Journal Royal Statistical Society*, B. 39: 1-38.

7. Dickey, D.A., and Fuller, W.A., 1979, Distribution of the estimations for autoregressive Time Series with a unit root, *Journal America Statistical Association*, 74: 427-431.

8. Fuller, W.A., 1995, *Introduction to Stationary Time Series* (2nd Edition), Wiley, New York.

9. Hurvic, C.M., and Tsai, C.L., 1989, Regression and Time Series model selection in small samples, *Biometrika*, 76: 297-307.

10. Quenouville, M.H., 1949, Approximate tests for correlation in Time Series, *Journal Royal Statistical Society*. B.11: 68-84.

11. Sahu, B.K., 1995, Statistical Inference from Geochemical and Petrographic Data, *Proceedings of Recent Research on Geology of Western India*, Baroda.

12. Schuster, A., 1906, On the periodicities of sunspots, *Phil. Trans. R. Soc. Sr.*, A.206:69-100.

13. Whittle, P., 1954, On stationary processes in the plane, *Biometrica*, 41: 434-449.

14. Winters, P.R., 1960, Forecasting sales by exponentially weighted moving averages, *Man. Sci.* 6: 324-342.

15. Akaike, H., 1969, Fitting autoregressive models for prodiction, Annals of the Inst. Statistical Mathematics, Tokyo, 21: 243-247.

16. Hurvich, C.M. and Tsai, C.L., 1989, Regression on Time Sereis model selection in small samples: *Biometrika*, 76: 297-307.

17. Anderson, O.D., 1976, *Time Series Analysis and Forecasting*, London, Butterworth, p. 182.

18. Wold, H. 1954, *A Study in the Analysis of Stationary Time Series*, Almquist & Wiksell, Stockholm.

19. Gronger, C.W.J., 1972, Time Series modelling and Interpretation. *European Econometric Congress*, Budapest, Sept. 72.

NON-STATIONARY UNIVARIATE TIME SERIES MODELS

3.1 INTRODUCTION

Many natural and experimental time series are non-stationary. They possess a certain homogeneity so that the random variable can be pre-transformed (differenced d times) before modelling in order to make the transformed series a stationary one. Then, the techniques of modelling and forecasting the transformed series proceed as per the methodologies developed in Chapter 2 for stationary sequences. The non-stationarity can be brought back by back-transformation (integration) of the trend, residual and forecasted values (Fig. 3.1).

There are two types of nonstationaries:

  (i) Stochastically-varying trend functions, which can be represented by a polynomial of (low) degree $d > 0$.

  (ii) Stochastically varying trend with some discontinuities at certain times.

The first type of non-stationarity can be modelled as an ARIMA(p, d, q) model, which means the d times differenced series is stationary with minimum variance, which has a model ARMA(p, q) fit. The second type of non-stationarity is usually handled by intervention analysis, i.e. the point of discontinuities are recognized and the data is homogeneous between two discontinuities and can be modelled as ARIMA(p, d, q). Intervention data arises because of: (i) policy changes in economics and social sciences, engineering production; and (ii) the presence of faults and unconformities, etc., in geological data. Positions of discontinuities may be recognized either by differencing the data or by using piecewise continuous cubic spline functions. Overdifferencing should be avoided as it then produces much larger residual (error) variance which, in turn, produces poor models and very poor forecasts. An AR model with unit root suggests the need for differencing and an MA model with unit root suggests that we have overdifferenced the series, which is to be avoided.

3.2 TRANSFORMATIONS FOR STATIONARITY

We assume that variance stabilizing transformation for X_t was modelled as

$$f_\lambda(X_t) = \begin{cases} \lambda^{-1}(X_t - 1) & X_t \geq 0, \lambda > 0 \\ \ln X_t, & X_t > 0, \lambda = 0 \end{cases} \quad (\text{Box–Cox Transform } f_\lambda)$$

Then, ARIMA(p, d, q) model may be written as

$$\zeta(B)x_i = \theta(B)z_i$$

where $\zeta(B)$ is a generalized autoregressive operator of degree $p + d$ with exactly d zeros (roots) equal to unity and all other roots outside the unit circle.

So,
$$\zeta(B) = \phi_p(B)(1 - B)^d = \phi_p(B) \nabla^d$$

where $\phi_p(B)$ is the stationary autoregressive operator of order p (as defined in Chapter 2) (Fig. 3.1, 3.3).

Differencing the x_i series d times and denoting the differenced series as w_i, i.e. $\nabla^d x_i = w_i$, we see that $\{w_i\}$ is ARMA(p, q) as the troubleshooting d roots of value 1.0 have been eliminated. Hence, the theory and methods of Chapter 2 would apply for modelling $\{w_i\}$ series. However, if $E(w_i) \neq 0 = \bar{w}$, then we subtract mean $\bar{w}$ for each data of w_i series in order to obtain $\tilde{w}_i = (w_i - \bar{w}_i)$, which is used for modelling (we delete ~ over w_i for convenience, henceforth). The non-stationarity of x_i is brought back by integrating or summing the $\{w_i\}$ process d times.

Suppose the current level of the process does not conform the behaviour of the time series (presence of discontinuity of constant value M). Then the model is

$$\zeta(B)(x_i + M) = \zeta(B)x_i$$

or
$$\zeta(B)M = 0$$

This implies $\zeta(1) = 0$, so $\zeta(B)$ has a factor $(1 - B)$ and if this is the only factor, then differencing the original series once ($d = 1$) will produce a stationary series for modelling as is required. If there are random shifts in slope, we need $\zeta(B)$ to have factor $(1-B)^2$ and second differencing $(d=2)$ is required to make the second differenced series stationary, and so on.

Usually, we restrict to $0 < d \leq 2$ for making these models easily explainable, as was done for modelling with $(p+q \leq 2)$.

Another method to recognize non-stationarity is through the use of its acf, which slowly decreases linearly with increasing lags. Multiplying the x_i series by x_{i-k} and taking expectations, we have for $k > q$:

$$\zeta(B)\rho_k = 0$$

The solution for $k > q - p - d$ is

$$\rho_k = A_1 + kA_2 + \cdots + k^{d-1}A_d + A_{d+1}\,\lambda_{d+1}^k + \cdots + A_{d+p}\,\lambda_{d+p}^k$$

where λ's are assumed to be distinct (for simplicity) and all having modulus < 1.

Since $|\rho_k| \leq 1$ as $k \to \infty$, we have $A_2 = A_d = 0$ and $\rho_k \to A_1$.

So, the acf tends to constant.

In particular for $\nabla^d x_i = z_i$, $\rho_k = 1$ for all k.

In practice, the series has finite length and it is impossible to know whether any of the AR zeros are exactly united. Even if some AR zero approaches 1, it has strong changes in level. Hence, it is best to difference the original series for stationarity, and then model the differenced series as AR which reduces the variance. Such series are called unstable series.

Suppose the length AR zero, $\lambda_r = 1 - \varepsilon$, where $\varepsilon > 0$ is distinct. Then, for moderate k,

$$\rho_k \to A_r \lambda_r^k \approx A_r(1 - k\varepsilon)$$

when ε is small.

Therefore, the acf will follow a gentle linear decline for an unstable series, in contrast to a rapid exponential decline for a stable series (ARMA(p, q)). Even if a few roots are coincident, the analyses can still be done. Finally, we have to sum/integrate the forecasted series d times.

This theoretical behaviour of a gentle linear decrease in ρ_k is also noticeable in sample acf $\{r_k\}$ series, which indicates that the series is unstable. The unstable behaviour is seen in pacf, which often has ϕ_{11} $(= r_1)$ very close to unity (Fig. 3.2).

If the series $\{x_t\}$ has a deterministic polynomial trend of degree d, we should replace w_i by $\tilde{w}_i = w_i - E[w_i]$ since $E[w_i] \neq 0$ for different t's. Writing the equation $\{x_t\}$ series as

$$x_i = f_0 + f_1 i + \cdots + f_d i^d + y_i$$

where $\phi(B)\nabla^d y_i = \theta(B) z_i.$

We have then $\nabla^d y_i = \phi^{-1}(B) \theta(B) z_i$

and $w_i = \nabla^d x_i = f_d \, d! + \phi^{-1}(B) \theta(B) z_i$

So, $\phi(B) w_i = \phi(1) f_d \, d! + \theta(B) z_i$

Taking expectations of both sides, we have

$$\phi(B) E[w_i] = \phi(1) f_d \, d!$$

So, $\phi(B) \tilde{w}_i = \theta(B) z_i.$

This shows that differencing removes both stochastic trends as well as deterministic trends. However, differencing also rapidly increases the noise variance, hence we must avoid overdifferencing at all costs.

Example 1

Compare the variances of x, ∇x, $\nabla^2 x$, where $\{x_i = i\lambda + z_i, i = 1, 2, \ldots N\}$, λ = fixed constant and $z_i \sim WN(0, \sigma^2)$.

Ans.

$$Var[x] = (\lambda^2(N^2 - 1)/12) + \sigma^2$$
$$Var[\nabla z] = 2\sigma^2 \text{(two times } \sigma^2)$$
$$Var[\nabla^2 z] = 6\sigma^2 \text{(six times } \sigma^2)$$

Example 2

Compare variance x_i, ∇x_i, $\nabla^2 x_i$, where $x_i = \sum_{j=1}^{j} J_j + z_i; i = 1, \ldots N$

and   (i) J_j, s are iid random variables with pdf $c(x) + \delta(x)$, where

$$c(x) = \int_{-\infty}^{\infty} c(x)dx = C \ll 1 \text{ and } \delta(x) \text{ dirac delta function.}$$

   (ii) $z_i \sim WN(0, \sigma^2)$ and are independent of $\{J_i\}$.
   (iii) there are $|J_i| \gg \sigma$.

This gives jumps in level at the time that dirac delta function operates. We have

$$Var[x_i] = i \, \sigma_j^2 + \sigma^2$$

$$Var[\nabla x_i] = \sigma_j^2 + 2\sigma^2 = Var[\nabla x]$$

$$Var[\nabla_2 x_i] = 2\sigma_j^2 + 6\sigma^2 = Var[\nabla^2 x]$$

This shows that overdifferencing doubles noise variance. So we should difference the above series only once, which becomes satisfactory for modelling purposes.

The eventual forecast function (e.f.f.) for ARIMA(p, d, q) model will be given by

$$\zeta(B)\,\hat{x}_i\,(l) = 0, l > q$$

with $p + d$ pivotal values $\hat{x}_i\,(q - p - d + 1)$, ..., $\hat{x}_i\,(q)$ (Fig. 3.4).

Example 3

Find e.f.f. for IMA(2,2) process

$$(1 - B)^2\,\hat{x}_i\,(l)\ = 0 \Rightarrow \hat{x}_i\,(l) = A + Cl, l > 0$$

with two pivotal values $\hat{x}_i\,(1) = A + C$, $\hat{x}_i\,(2) = A + 2C$.

Hence e.f.f. is

$$\hat{x}_i\,(l)\ = \{2\,\hat{x}_i\,(1) - \hat{x}_i\,(2)\} + \{\,\hat{x}_i\,(2) - \hat{x}_i\,(1)\}l,$$

a straight line through points $\hat{x}_i\,(1)$ and $\hat{x}_i\,(2)$, which is given by

$$\hat{x}_i\,(1)\ = 2x_i - x_{i-1} + \theta_1 z_i + \theta_2 z_{i-1}$$

$$\hat{x}_i\,(2)\ = 2\,\hat{x}_i\,(1) - x_i + \theta_2 z_i.$$

Example 4

Find e.f.f. for ARIMA(1,1,1) process

$$(1 - B)\,(1 - \phi B)\,\hat{x}_i\,(l) = 0 \Rightarrow \hat{x}_i\,(l) = A + C\phi^l, l > -2$$

with pivotal values $\hat{x}_i\,(-1)\ = x_{i-1} = A + \dfrac{C}{\phi}$

and $\quad x_i(0) = x_i = A + C.$

Hence, $\quad A = (x_i - \cdots - \phi x_{i-1})/(1 - \phi); C = -\dfrac{\phi\,(x_i - x_{i-1})}{(1 - \phi)}$

Example 5

Find e.f.f. for IMA(1,1)

$$x_i\ = x_{i-1} + z_i + \theta z_{i-1}$$

and its e.f.f. is $\hat{x}_i\,(1) = x_i + \theta z_i$

$$\hat{x}_i(1) = \hat{x}_i(l-1) \text{ for all } l > 1.$$

So, forecasts made for all leads are same as the one-step ahead forecasts

$$\hat{x}_i(1) = x_i = \theta z_i$$

or
$$\hat{x}_i(1) = (1 + \theta)x_i - \theta(x_i - z_i)$$

$$= (1 + \theta)x_i - \theta(x_{i-1} - \theta z_{i-1})$$

$$= (1 + \theta)x_i - \theta\hat{x}_{i-1}(1)$$

This is also called exponential smoothing recursive formula for updating one-step ahead forecasts. Successive $\hat{x}_i(1)$'s are thus obtained by interpolating between the previous forecast and the current value.

By successively eliminating $\hat{x}_{i-j}(1); j = 1, 2, ...,$ we get

$$\hat{x}_i(1) = (1 + \theta)\sum_{j=0}^{\infty}(-\theta)^j x_{i-j}$$

Since $|\theta| < 1$ for invertibility, this is an exponentially weighted moving average (EWMA) of previous values. So, IMA(1,1) is also called a EWMA process.

For ARIMA(p, 1, q), forecast levels must also be continuously updated and for ARIMA(p, 2, q), we should continuously update both levels as well as the slopes. Therefore, if the main purpose is forecasting, it would be better to use an ARIMA model rather than an ARMA one.

Example 6

Find variance $V(l)$ for AR(1), ARMA(1,1), IMA(1,1)

AR(1) $\psi(B) = (1 - \phi B)^{-1} = 1 + \phi B + \phi^2 B^2 + \cdots$

So, $V(l) = \left(\dfrac{1 - \phi^{2l}}{1 - \phi^2}\right)\sigma^2 \rightarrow \sigma^2/(1 - \phi^2)$

ARMA(1,1) $\psi(B) = 1 + (\phi + \theta)B + \phi(\phi + \theta)B^2 + \phi^2(\phi + \theta)B^3 + \cdots$

So, $V(l) = \left[1 + (\phi + \theta)^2\left(\dfrac{1 - 2^{2l-2}}{1 - \phi^2}\right)\right]\sigma^2 \rightarrow \left(\dfrac{1 + 2\phi\theta + \theta^2}{1 - \phi^2}\right)\sigma^2$

IMA(1,1) $\psi(B) = 1 + (1 + \theta)B + (1 + \theta)B^2 + \cdots$

So, $V(l) = [1 + (l - 1)(1 + \theta)^2]\sigma^2$

In addition, we could also use weighted linear combinations of the different available forecast functions as a final forecasting procedure, which has been advocated by Granger and his associates.

Example 7

Find acf and pacf of ARIMA(1,1,0) with $|\phi| < 1$.

We have $(1 - \phi B)(1 - B)X_t = z_t; \{z_t\} \sim WN(0, \sigma^2)$.

So,
$$X_t = X_0 + \sum_{j=1}^{\infty} \phi^j z_{t-j} \, .$$

Here, the acf decreases very slowly and the linear suggests the need for differencing to make the series stationary. Only two pacf values are significant at the 95% level, so we may fit an AR(2) model to this undifferenced data. Assuming $EX_t = 0$, we obtain the best fit AR(2) model as

$$(1 - 1.808B + .811B^2)X_t = (1 - .825B)(1 - .983B)X_t = z_t$$

with $\{z_t\} \sim WN(0, .970)$ which is stationary but the coefficient $\phi_2 = .983$ is very close to unit root 1.0, hence once differencing seems necessary.

An obvious limitation of ARIMA(p, d, q) model is that we permit non-stationarity only in a very special way, i.e. by allowing polynomial $\phi(B)X_t = \theta(B)z_t$ to have a zero of multiplicity d at the point 1 on the unit circle. Similarly, if autocorrelations decay slowly but have oscillatory behaviour then $\zeta(B)$ has a zero close to e^{iw} for some $w \in (-\pi, \pi]$ other than 0.

Suppose $X_t - (2r^{-1} \cos w)X_{t-1} + r^2 X_{t-2} = z_t, \{z_t\} \sim WN(0, 1)$

with $r = 1.005$ and $w = \pi/3$, i.e.

$$X_t - .9950X_{t-1} + .9901X_{t-2} = z_t, \{z_t\} \sim WN(0,1).$$

The acf can be obtained by noting that

$$1 - (2r^{-1} \cos w)B + r^{-2}B^2 = (1 - r^{-1} e^{iw} B)(1 - r^{-1} e^{-iw}B)$$

and $\quad \rho(h) = r^{-h} \dfrac{\sin(hw + \psi)}{\sin \psi}, h \geq 0$ (causal AR(2) process)

where $\quad \tan^2 \psi = \dfrac{r^2 + 1}{r^2 - 1} \tan w.$

It is clear now that $\rho(h) \to \cos(hw)$ as $r \downarrow 1$.

So, acf is a damped sine wave with damping ratio $1/1.005$ and period 6.

However, if we take $r = 1$ and $w = \pi/3$, then the AR polynomial becomes $(1 - B + B^2)$.

On the other hand, if we use $2\pi/3 = w =$ some integer s (in this case $s = 6$), then the series is seasonal and we can use operator $(1 - B^6)$ to make the series non-seasonal ARMA. In both these cases, $\phi(B)$ zeros will lie outside the unit circle and ARMA model can be easily identified (Fig. 3.4).

3.3   UNIT ROOT PROBLEMS

The model may have a root in or near the unit circle for AR and/or MA polynomials. If AR polynomial has a unit root, the data should be differenced before the ARMA model is fitted. If MA polynomial has a unit root, then the data has been overdifferenced and hence a lesser value of d is suggested.

If $\{X_t\}$ belongs to AR(1) model, i.e.

$$X_t - \mu = \phi(X_{t-1} - \mu) + z_t; \; \{z_t\} \sim WN(0, \sigma^2)$$

with $|\phi| < 1$, $\mu = EX_t$.

For large n, the MLE $\hat{\phi}_1$ of ϕ_1 is approximately $N(\phi_1, (1 - \phi_1^2)/n)$, but this is not valid distribution if $\phi = 1$.

We can make H_0 test by differencing to obtain $\nabla X_t = X_t - X_{t-1} - 1$.

Let $\hat{\phi}_{OLS}^*$ estimator of $\hat{\phi}_1^*$ by regressing ∇X_t on 1 and X_{t-1}.

The standard error of $\hat{\phi}_1^*$ is given by

$$S\hat{E}(\hat{\phi}_1^*) = S \Big/ \left(\sum\nolimits_{t=2}^{n} (X_{t-1} - \overline{X})^2 \right)^{1/2}$$

where $S^2 = \sum\nolimits_{t=2}^{n} (\nabla X_t - \hat{\phi}_0^* - \hat{\phi}_1^* X_{t-1})^2 /(n - 3)$ and $\overline{X}$ is sample mean.

As $n \to \infty$, the t ratio limit distribution is

$$\hat{\tau}_\mu = \hat{\phi}_1^* / S\hat{E}(\hat{\phi}_1^*)$$

where $\hat{\phi}_1^* = 0$ and we can contract test of null hypothesis $H_0: \phi_1 = 1$.

τ_μ values for .01, .05 and .01 quantiles of the limit distribution $\hat{\tau}_\mu$ are -3.43, -2.86, and -2.57, respectively. Null hypothesis of unit root is rejected if $\hat{\tau}_\mu < -2.86$.

The same procedure is extended is for AR(p) processes.

The unit root in MA polynomial can have a number of interpretations, depending on the modelling application.

If $\{X_t\}$ is causal invertible ARMA(p, q) process

$$\phi(B)X_t = \theta(B)z_t, \{z_t\} \sim WN(0, \sigma^2).$$

Then the differenced series $Y_t = \nabla X_t$ is a non-invertible ARMA($p, q + 1$) process with MA polynomial $\theta(z)(1 - z)$. So, testing the unit root of MA polynomial is equivalent to testing for overdifferencing (i.e. variance of differenced series increases from minimum variance at d).

We can distinguish between two competing models thus:

$$\nabla^k X_t = a + V_t \text{ (no MA unit roots)}$$

$$X_t = c_0 + c_1 t + \cdots + c_k t^k + W_t \text{ (multiple MA with roots of order } k).$$

where $\{V_t\}$ and $\{W_t\}$are invertible ARMA processes.

The following example is given to illustrate the problem of unit root and how it can be eliminated.

For a MA(1) process $X_t = z_t + \theta z_{t-1}, \{z_t\} \sim IID(0, \sigma^2)$.

If H_0: $\theta = -1$ and H_1: $\theta > -1$ can be tested as $n(\hat{\theta} + 1)$ converges in distribution.

We reject H_0 whenever $\hat{\theta} > -1 + c_\alpha/n$

where c_α is $(1 - \alpha)$-quantile limit distribution of $n(\hat{\theta} + 1)$.

$c_{.01} = 11.93$, $c_{.05} = 6.80$, $c_{.10} = 4.90$.

If $n = 50$, H_0 is rejected at 0.05 level if $\hat{\theta} > -1 + 6.80/50 = -.864$.

A likelihood ratio test can also be used for testing the unit root hypothesis.

The LR for this problem is $L(-1, S(-1)/n)/L(\hat{\theta}, \hat{\sigma}^2)$.

The model hypothesis is rejected at level α if

$$\lambda_n: = -2 \ln \left(\frac{L(-1, S(-1)/n)}{L(\hat{\theta}, \hat{\phi}^2)} \right) > c_{LR, \alpha},$$

where $P_{\theta=-1}|\lambda_n > c_{LR,\alpha}| = \alpha$.

If $n \geq 50$, the limiting quantities for λ_n under H_0 are

$c_{LR,0.01} = 4.41$

$$c_{LR,0.05} = 1.94$$

$$c_{LR,0.10} = 1.00$$

3.4   FORECASTING ARIMA MODELS

If $d \geq 1$, the first and second moments EX_t and $E(X_{t+h} X_t)$ are not determined by the difference equations, so we cannot expect to determine the best linear predictors for $\{X_t\}$ without making further assumptions. Suppose $\{Y_t\}$ is a causal ARMA(p, q) and X_0 is a random variable. We define X_t as

$$X_t = X_0 + \sum_{j=1}^{t} Y_t, t = 1, 2, \ldots$$

Then $\{X_t, t \geq 0\}$ is an ARIMA(p, 1, q) process with mean $EX_t = EX_0$ and autocovariance $E(X_{t+h} X_t) - (EX_0)^2$ that depends upon Var(X_0) and Cov (X_0, Y_j), $j = 1, 2, \ldots$

The best linear predictor X_{n+1} based on $\{1, X_0, X_1, \ldots, X_n\}$ is the same as the best linear predictor in terms of the set $\{1, X_0, Y_1, \ldots, Y_n\}$ since both X_t and Y_t are linear combinations. Hence, deviations P_n to be the best linear predictor for both the sets and using linearity of P_n, we have $P_n X_{n+1}$ $= P_n(X_0 + Y_1 + \cdots + Y_{n+1}) = P_n(X_n + Y_{n+1}) = X_n + P_n Y_{n+1}$.

To evaluate $P_n Y_{n+1}$, it is necessary to know $E(X_0, Y_j), j = 1, 2, \ldots, n + 1$ and EX_0^2. But assuming X_0 is uncorrelated with $\{Y_t, t \geq 0\}$, then $P_n Y_{n+1}$ is the best linear predictor of $\hat{Y}_{n+1}$ of Y_{n+1} in terms of the set $\{1, Y_1, \ldots, Y_n\}$, which can be calculated as in the case of ARMA(p, q). Therefore, the assumption that X_0 is uncorrelated with $Y_1, \ldots, Y_n$ is sufficient to determine the best linear predictor $P_n Y_{n+1}$ in this case.

However, in the general case of ARIMA(p, d, q), we have

$$(1 - B)^d X_t = Y_t, t = 1, 2, \ldots.$$

where $\{Y_t\}$ is causal ARMA(p, q) process, and the random vector $(X_{t-d}, \ldots, X_0)$ is uncorrelated with $Y_t, t > 0$.
The difference equation can be rewritten as

$$X_t = Y_t - \sum_{j=1}^{d} \binom{d}{j}(-1)^j X_{t-j}, t = 1, 2, \ldots \qquad \ldots(A)$$

It is convenient to relabel the time axis to assume that we observed $X_{1-d}, X_{2-d}, \ldots, X_n$ and observed values of $\{Y_t\}$ are then $Y_1, \ldots, Y_n$ as before. If

P_n is best linear predictor of observations up to time n, i.e. 1, X_{1-d}, X_{2-d}, ...,
X_n or equivalently 1, X_{1-d}, X_{2-d}, ..., X_0, Y_1, ..., Y_n. We wish now to compute
$P_n X_{n+h}$. We apply P_n to both sides of the equation (A) (with $t = n + h$) and
using linearity of P_n, we obtain

$$P_n X_{n+h} = P_n Y_{n+h} - \sum_{j=1}^{d} \binom{d}{j} (-1)^j P_n X_{n+h-j} \qquad \qquad ...(B)$$

Since $(X_{1-d}, X_{2-d}, ..., X_0)$ is uncorrelated with Y_t, $t > 0$, we can find
$P_n Y_{n+h}$ as the best linear predictor of Y_{n+h} in terms of $\{1, Y_1, ..., Y_n\}$ and
this is computed as in ARMA(p, q) forecasting. We note $P_n X_{n+1-j} = X_{n+1-j}$ for each $j \geq 1$ and hence $P_n X_{n+1}$ is obtained for equation (B). Next, we
compute $P_n X_{n+2}$ and recursively $P_n X_{n+3}$, $P_n X_{n+4}$, etc.

To find MSE of predictors, it is convenient to express $P_n Y_{n+h}$ in terms
of $\{X_j\}$.

For $n \geq 0$, we denote the one-step predictors by $\hat{Y}_{n+1} = P_n Y_{n+1}$ and

$$\hat{X}_{n+1} = P_n X_{n+1}.$$

Then, $\qquad \qquad X_{n+1} - \hat{X}_{n+1} = Y_{n+1} - \hat{Y}_{n+1}, n \geq 1$

and if $n > m = \max(p, q)$ and $h \geq 1$, we write

$$P_n Y_{n+h} = \sum_{i=1}^{p} \phi_i P_n Y_{n+h-i} + \sum_{j=1}^{q} \theta_{n+h-1,j} (X_{n+h-j} - \hat{X}_{n+h-j})$$

Setting $\zeta(B) = (1 - B)^d \phi(B) = 1 - \zeta_1 B - \cdots - \zeta_{p+d} B^{p+d}$, we obtain

$$PX_{n+1} = \sum_{j=1}^{p+d} \zeta_j P_n X_{n+h-j} + \sum_{j=h}^{q} \theta_{n+h-1,j} (X_{n+h-j} - \hat{X}_{n+h-j})$$

which is analogous to h-step prediction formula for ARMA(p, q) process.
The MSE for h-step predictor in ARIMA(p, d, q) model is

$$\sigma_n^2 (h) = E(X_{n+h} - PX_{n+h})^2 = \sum_{j=0}^{h-1} \left(\sum_{r=0}^{j} \chi_r \theta_{n-h+r-1, j-r} \right)^2 v_{n+h-j-1}$$

where $\theta_{n_0} = 1$, $\chi(B) = \sum_{r=0}^{\infty} \chi_r B^r (1 - \zeta_1 B - \cdots - \zeta_{p+d} B^{p+d})^{-1}$

and $v_{n+h-j-1} = E(X_{n+h-j} - \hat{X}_{n+h-j})^2 = E(Y_{n+h-j} - \hat{Y}_{n+h-j})^2$

For large $n \to \infty$, provided $\theta(\cdot)$ is invertible, we obtain

$$\sigma_n^2(h) = \sum_{j=0}^{h-1} \psi_j^2 \, \sigma^2$$

where $\psi(B) = \sum_{j=0}^{\infty} \psi_j B^j = (\zeta(B))^{-1} \theta(B)$.

Once the forecast function weights $a_0, a_1, ..., a_d, b_1, ..., b_p$ are determined for the forecast function $\hat{x}_{n+h} = a_0 + a_1 h + \cdots + a_d h^{d-1} + b_1 \xi_1^{-h}$

$+ \cdots + b_p \xi_p^{-h}$, using $p + d$ equations for $q - p - d < h \leq q$, we can compute predictors for all values of $h > q_j$, where the ξ's are roots (zeros) of the $\phi_p(B)$ assumed to be distinct (see Figs. 3.5, 3.6).

3.5 SARIMA MODELS

If $\{X_t\}$ has a seasonality s, then differencing by operator $(1 - B^s)$, we obtain Y_t which is non-seasonal and can be modelled as ARIMA(p, d, q).

Thus, $Y_t = (1 - B^s)X_t$ and $\phi_p(1 - B^s)X_t = (B)z_t$

If d and D are non-negative integers, then $\{X_t\}$ is SARIMA $(p, d, q) \times (P, D, Q)$ process with period s if the differenced series $Y_t = (1 - B)^d (1 - B^s)^D X_t$ is a causal. ARMA process defined by

$$\phi_p(B)\,\Phi_P(B^s)Y_t = \theta_q(B)\Theta_Q(B^s)z_t, \; \{z_t\} \sim WN(0, \sigma^2).$$

The process $\{Y_t\}$ is causal if $\phi_p(B) \neq 0$, $\Phi_P(B) \neq 0$ for $|B| \leq 1$.

In most applications, D is rarely more than one and P and Q are normally less than three. Using the difference process $\{Y_t\}$, the model becomes

$$\phi_{p+sP}(B)Y_t = \theta_{q+sQ}(B)z_t$$

Provided $p < s$ and $q < s$, the constraints in the coefficients of $\phi_{p+sP}(B)$ and $\theta_{q+sQ}(B)$ can all be expressed as multiplicative relations

$$\phi_{is+j} = \phi_{is}\phi_j, \; i = 1, 2, ...; j = 1, 2, ... s - 1$$

$$\theta_{is+j} = \theta_{is}\theta_j, \; i = 1, 2, ...; j = 1, 2, ... s - 1$$

In contrast to the classical decomposition of additive seasonal models where s_t repeats exactly from cycle to cycle, SARIMA models are multiplicative and allow randomness in seasonal pattern from one cycle to the next.

Example 8

Forecasting for ARIMA(1,1,0) model

The Dow-Jones Utilizes Index (Aug. 28 – Dec. 18, 1972) gives an AR(1) model for mean-corrected differences denoted as X_t.

The model was

$$X_t - .4471X_{t-1} = z_t, \{z_t\} \sim WN(0,0.1455)$$

where $X_t = D_t - D_{t-1} - .1336, t = 1, ..., 78$ and $\{D_t, i = 1, ..., 77\}$ is the original series.

The model for $\{D_t\}$ thus becomes

$$(1 - .4471B)(1 - B)D_t = z_t + .0739, \{z_t\} \sim WN(0,0.1455).$$

The forecast function is given by

$$P_nD_{n+h} = P_nX_{n+h} + .1336 + P_nD_{n+h-1} = (.4471)^hX_n + .1336 + P_nD_{n+h-1}$$

where P_nD_{n+h} is the best linear predictors of the zero-mean AR(1) process X_t.

To compute one- and two-step predictors for pivot 77, we get

$$P_{77}D_{78} = (.4471)(-.9036) + .1336 + 121.23 = 120.960$$

$$P_{77}D_{79} = (.4471)^2(-.9036) + .1336 + 120.960 = 120.913$$

The corresponding variances for forecast functions are

$$\sigma_{77}^2(1) = \sigma^2 = 0.1455$$

$$\sigma_{77}^2(2) = \sigma^2(1 + 1.4471)^2 = .4502$$

Suppose we have monthly time series data for several years and each month data have the same ARMA(P, Q) model, so that each month data can be regarded as a separate realization of the same process. Thus, the model can be considered as the between-year model, written completely as: $\phi_P(B^{12})Y_t = \Theta_Q(B^{12})U_t$

with $\{U_t\} \sim WN(0, \sigma_U^2)$.

However, $E(U_tU_{t+h})$ is not necessarily zero except when h is an integer multiple of 12. Therefore, U_t may be modelled as an ARMA(p, q) as follows:

$$\phi_p(B)U_t = \theta_q(B)z_t, \{z_t\} \sim WN(0, \sigma^2).$$

This assumption implies a non-zero correlation between the consecutive values of U_t but also within the 12-monthly sequences $\{U_{j+12}, t = ..., -1, 0, 1, ...\}$, and $\{U_t\}$ is not $WN(0, \sigma_U^2)$.

The first step for a SARIMA model is to identify d and D, so as to make these differenced observations $Y_t = (1 - B)^d (1 - B^s)^D X_t$ stationary.

Next, we can compute acf and pacf of $\{Y_t\}$ at lags, which are multiples of s, in order to find P and Q.

If $\hat{\rho}(\cdot)$ is sample acf of $\{Y_t\}$, then P and Q are chosen so that $\hat{\zeta}(ks)$, $k = 1, 2, ...$ is compatible with acf of an ARMA(P, Q) process. The orders p, q is then selected by trying to match $\hat{\rho}(1), ..., \hat{\rho}(s - 1)$ with the acf of an ARMA(p, q) process. Finally, AICC and goodness-of-fits have to be made for selecting best SARIMA model among the different candidates.

For given values of $p, d, q, P, D,$ and Q, the parameters θ, Φ, Θ and σ^2 can be found using the maximum likelihood method. The differencing $Y_t = (1 - B)^d (1 - B^s)^D X_t$ constitute an ARMA$(p + sP, q + sQ)$ processes in which some coefficients may be zero and the rest are functions of the $(p + P + q + Q)$ dimensional vector

$$\beta' = (\phi', \Phi', \theta', \Theta')$$

For fixed β, the reduced likelihood $l(\beta)$ of the differences $Y_{t + td + sd}$ is easily computable. A more direct approach to model the differenced series $\{Y_t\}$ is simply to fit a subset ARMA model without making use of the multiplicative form.

Forecasting is easy and we have

$$X_{n+h} = Y_{n+h} + \sum_{j=1}^{d+Ds} a_j X_{n+h-j}$$

Assuming first $d + Ds$ observations with $X_{-d-Ds+1}, ..., X_0$ are uncorrelated with $\{Y_t, t \geq 1\}$, we can obtain best linear predictors $P_n X_{n+h}$ of X_{n+h} based on $\{1, X_{-d-Ds+1}, ..., X_n\}$ by applying P_n to each side to yield

$$P_n X_{n+h} = P_n Y_{n+h} + \sum_{j=1}^{d+Ds} a_j P_n X_{n+h-j}$$

where $P_n Y_{n+h}$ is the non-zero mean ARMA.

The predictors $P_n X_{n+h}$ can then be computed recursively for $h = 1, 2, ...,$ noting that $P_n X_{n+1-j} = X_{n+1-j}$ for each $j \geq 1$.

For large n,

$$\sigma_n^2(h) = E(X_{n+h} - P_n X_{n+h})^2 = \sum_{j=0}^{h-1} \psi_j^2 \sigma^2$$

where $\psi(B) = \sum_{j=0}^{\infty} \psi_j B^j = \dfrac{\theta(B)\Theta(B^s)}{\phi(B)\Phi(B^s)(-B)^d(1 - B^s)^D}$, $|B| < 1$.

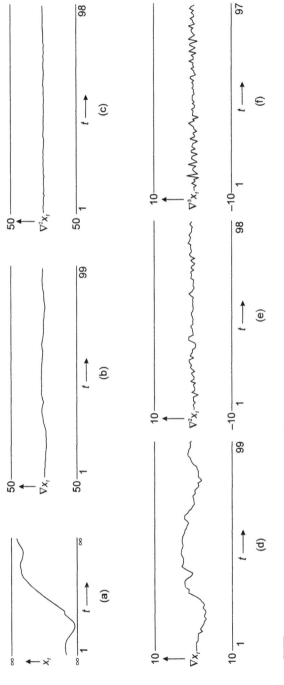

Fig. 3.1 Two sets of simulated time series $(1 - B)^2 x_t = z_t$ with its first, second and third series (which shows that the second differenced series produces stationarity)

Note: Series (d) and (e) are magnified images of series (b) and (c), respectively.

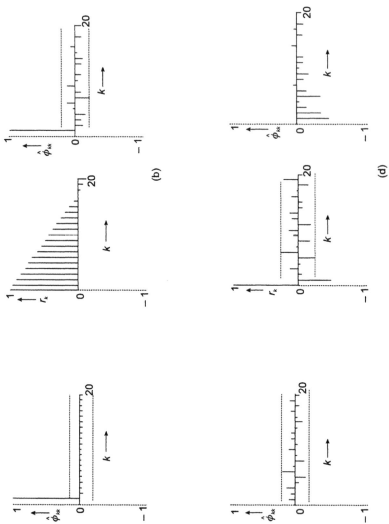

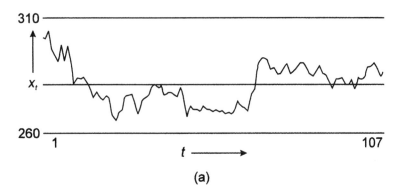

(a)

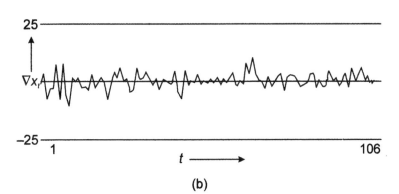

(b)

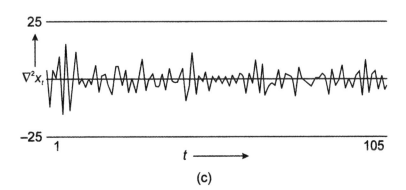

(c)

Fig. 3.3 A typical closing stock prices data with first and second differenced series

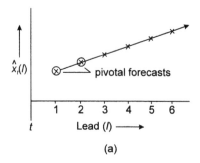

(a)

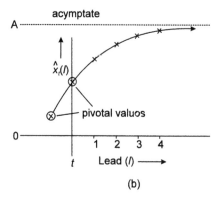

(b)

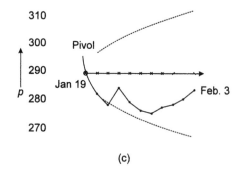

(c)

Fig. 3.4   Effective forecast function (e.f.f.) for
(a) IMA(2,2)
(b) ARIMA(1,1,0)
(c) Random walk fit to ICI stock price

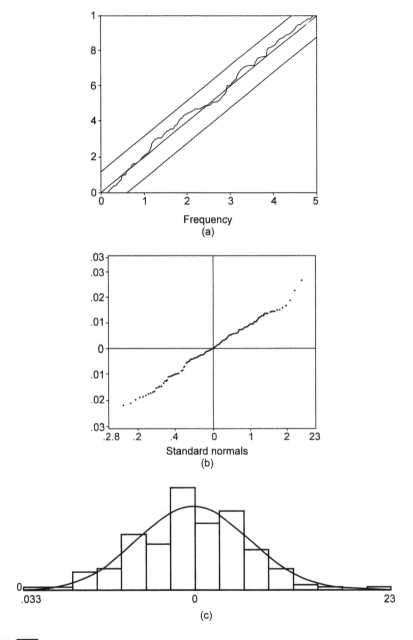

Fig. 3.5 Diagnostic check for residuals of the time series through
(a) Cumulative spectral distribution
(b) Q-Q plot and
(c) Histogram showing normal distribution for the residuals

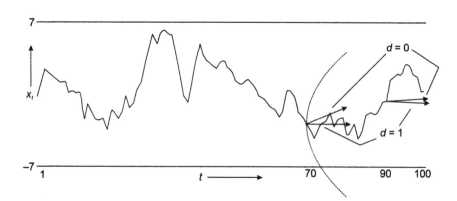

Fig. 3.6 Time series showing forecast functions at different pivots assuming the series to be stationary ($d = 0$)

3.6 EXERCISES

1. If $\{X_t\}$ is ARIMA(p, d, q) process with $\{z_t\} \sim WN(0, \sigma^2)$, then show that the difference equations are also satisfied by the process

$$W_t = A_0 + A_1 + \cdots + A_d t^{d-1}$$

   where $A_d, i \in (0, d-1)$ are arbitrary random variables.

2. Let $\{X_t\}$ be ARIMA(2,1,0) process with $\phi_1 = .8$, $\phi_2 = -.25$ and $\{z_t\} \sim WN(0,1)$.

   Find forecast function $P_n X_{n+h}$ for $h > 0$.

   Assume n to be large. Then compute $\sigma_n^2(h)$ for $h = 1$ to 5.

3. Let $\{X_t\}$ be differenced at lag 12, then at lag 1 to produce a zero-mean series $\{Y_t\}$ with following acf:

   $\hat{\rho}(12j) \approx (.8)^j$; $j = 0, \pm 1, \pm 2, \ldots$

   $\hat{\rho}(12j \pm 1) \approx (.4)(.8)^j$; $j = 0, \pm 1, \pm 2, \ldots$

   $\hat{\rho}(h) \approx 0$;   otherwise

   and $\hat{\gamma}(0) = 25$.

   (a) What is the SARIMA(P, D, Q) model for $\{X_t\}$?

   (b) For large n, find one- and twelve-step linear predictors $P_n X_{n+1}$ and $P_n X_{n+1}$ in terms of X_t, $t = -12, -11, \ldots, n$ and for $Y_t - \hat{Y}_t$, $t = 1, \ldots, n$.

   (c) Find the mean-square errors of these predictors.

4. For SARIMA$(1,0,0)$ $(0,1,1)_{12}$ model

$X_t = X_{t-12} + \phi(X_{t-1} - X_{t-13}) + z_t + \theta z_{t-12}$

show that best linear forecast $x(N, 2) = x_{N-0} + \phi^2(x_N - x_{N-2})$ $+ \theta\phi z_{N-11} + \theta z_{N-10}$.

5. Consider ARIMA$(0,1,1)$ process $(1 - B)X_t = (1 - \theta B)z_t$

Show that best linear forecast $\hat{x}(N, 1) = x_N - \theta z_N$ and $\hat{x}(N, k)$ $= \hat{x}(N, k - 1)$ for $k \geq 2$.

Show that $\hat{x}(N, 1)$ is equivalent to exponential smoothing.

Prove the variance of k-step ahead forecast function error is $[1 + (k - 1) (1 - \theta)^2]\sigma_z^2$.

6. Give the solution to the superposition of ARIMA (p_1, d_1, q_1) and ARIMA (p_2, d_2, q_2), where each $p_i, d_i, q_i, i = 1, 2$ are non-zero and $d_1 \geq d_2$ [Hint: Use Granger's lemma].

Ans: Superposed model is ARIMA(p, d, q), where $p = p_1^* + p_2^* + h$,

$d = d_1, q \leq \max [p_2^* + q_1, p_1^* + d_1 - d_2 = q_2]$.

Here, $\phi_{p_1}(B) = h(B) \phi_{p_1}(B)$ and $\phi_{p_2}(B) = h(B) \phi_{p_2}(B)$.

References

1. Anderson, O.D., 1976, *Time Series Analysis and Forecasting:* Butterworth, London, 182 pp.
2. Box, G.E.P. and Cox, D.R., 1964. An Analysis of Transformations, *Journal Royal Statistical Society*, B26: 211-252.
3. Box, G.E.P., Jenkins, G.M. and Reinsel, 1994, *Time Series Analysis: Forecasting and Control*, Prentice Hall, Englewood Cliffs, NJ (3rd Edition).
4. Brockwell, P.J. and Davis, R.J., 1991, *Time Series: Theory and Methods* (2nd Edition), Spinger, New York.
5. Dempster, A.P., Laird, N.M. and Rubin, D.B., 1977, Maximum likelihood for incomplete data via EM algorithm, *Journal Royal Statistical Society*, B: 39: 1-38.
6. Quenouille, M.H., 1949, Approximate tests for correlation in Time Series, Journal Royal Statistical Society, B11: 68-65.
7. Frances, H.P., 1999, *Time Series Models for Business and Economics Forecasting*, Cambridge University Press, UK, 280 pp.
8. Lawrence, A.J., 1991, Directionality and Reversibility in Time Series, *Institute of Statistical, Review*, 59: 67-79.
9. Parzen, E. (Ed.), 1989, Time Series Analysis of Irregularly Observed Data, Lecture Notes. Vol. 25, Springer, New York.
10. Priestley, M.B., 1981, *Spectral Analysis and Time Series*, Academic Press, London.
11. Priestley, M.B., 1988, *Non-linear and Non-stationary Time Series Analysis*, Academic Press, London.
12. Ripley, B.D., 1981, *Spatial Statistics*, Wiley, Chichester.
13. Tong, H., 1990, *Non-linear Time Series*, Clarendon Press, Oxford.
14. Young, P.C., 1984, *Recursive Estimation and Time Series Analysis*, Springer, Berlin.

VECTOR AND MULTIDIMENSIONAL TIME SERIES ANALYSIS

4.1 INTRODUCTION

Many time series are best considered as components of some vector-valued (multivariate/multidimensional) time series $\{X_t\}$ having serial dependence not only within each component series $\{X_{ti}\}$, but also interdependence between different component series $\{X_{ti}\}$ and $\{X_{tj}\}$, $i \neq j$. Much of the theory of univariate time series can be extended to multivariate time series but interdependence between two time series has to be considered in addition as cross-covariance/cross-correlation. First, we consider the basic properties of stationary multivariate time series having multivariate white noise and estimation of mean vector, covariance matrices at different lags and testing serial independence based on multivariate observations. Next, we consider the multivariate ARMA models (which are linear) and their identification which is simplified by restricting our attention to vector autoregressive (VAR) processes. We can also model by using multivariate Yule-Walker equations and by generalizing Durbin-Levison algorithm. Forecasting multivariate series may be made by using known second-order properties as also transfer functions. We also extend our analysis to cover non-linear and nonstationary multivariate time series models and give suitable examples for earth sciences as well as for other disciplines.

In order to motivate, we first introduce two bivariate time series. A bivariate time series is a series of two-dimensional vectors (X_{t1}, X_{t2}) observed at time t (usually $t = 1, 2, 3, ...$), which is common in earth sciences as map analysis. Here, we should analyze the two series X_{t1} and X_{t2} for their autocovariances (autocorrelations) as well as cross-correlations between X_{t1}, X_{t2}, which is of importance in forecasting.

If we have m time series $\{X_{ti}\}$, $i = 1, 2, ... m$, we have convenient vector notation as

$$X_t = \begin{bmatrix} X_{t1} \\ X_{t1} \end{bmatrix}, t = 0, \pm 1, ...$$

The second-order properties of multivariate time series $\{X_t\}$ are specified by mean vectors

$$\mu_t = EX_t = [\mu_{t1} ...... \mu_{tm}]'$$

and covariance matrices

$$\Gamma(t + h, h) = \begin{bmatrix} \gamma_{11}(t + h, t) & ...... & \gamma_{1m}(t + h, t) \\ \gamma_{m1}(t + h, t) & ...... & \gamma_{mm}(t + h, t) \end{bmatrix}$$

where $\gamma_{ij}(t + h, t) = Cov(X_{t+h,i}, X_{t,j})$.
In matrix notation $\Gamma(t + h, t) = E(X_{t+h} - \mu_{t+h})(X_t - \mu_t)$.

If X_{ti} are multivariate normal with $EX_{ti}^2 < \infty$ for all t and i, then the mean vectors and covariance matrices are sufficient to specify all joint distributions. Even if observations $\{X_{ti}\}$ do not have joint multivariate distributions, mean vector and covariance matrices provide a measure of dependence (correlation) not only between observations of the same series, but also between the same in different series. The m-variate series $\{X_t\}$ is weakly stationary if the following holds:

$\mu_X(t)$ is independent of t and $\Gamma(t + h, t)$ is independent of t for each h.
Thus, we have for a stationary m-variate time series

$$\mu = EX_t = [\mu_1 ... \mu_m]'$$

and $$\Gamma(h) = E[(X_{t+h} - \mu)(X_t - \mu)'] = \begin{bmatrix} \gamma_{11}(h) & ... & \gamma_{1m}(h) \\ ... & ... & ... \\ ... & ... & ... \\ \gamma_{m1}(h) & ... & \gamma_{mm}(h) \end{bmatrix}$$

We note that $\gamma_{ij}(\cdot) \neq \gamma_{ji}(\cdot)$, i.e. it is an asymmetric function.
The correlation matrix function is

$$R(h) = \begin{bmatrix} \rho_{11}(h) & \cdots & \rho_{1m}(h) \\ \cdots & \cdots & \cdots \\ \cdots & \cdots & \cdots \\ \rho_{m1}(h) & \cdots & \rho_{mm}(h) \end{bmatrix}$$

where $\rho_{ij}(h) = \gamma_{ij}(h) / [\gamma_{ii}(0)\gamma_{ij}(0)]^{1/2}$.

Basic properties of covariance matrix $\Gamma(\cdot)$ are:

(i) $\Gamma(h) = \Gamma'(-h)$

(ii) $|\gamma_{ij}(h)| \le [\gamma_{ii}(0)\gamma_{jj}(0)]^{1/2}; i, j = 1, \ldots m$

(iii) $\gamma_{ii}(\cdot)$ is an autocovariance function $i = 1, \ldots, m$

(iv) $\sum_{jjk=1}^{n} a'_j \Gamma(j - k)a_k \ge 0$ for all $n \in \{1, 2 \ldots\}$ and $a_1 \ldots a_n \in R^m$.

The correlation matrix $R(\cdot)$ also have the above properties and, in addition, $\rho_{ii}(0) = 1$ for all i. The correlation $\rho_{ij}(0)$ between X_{ti} and X_{tj} is not generally equal to 1 if $i \ne j$ and also $|\gamma_{ij}(h)| > |\gamma_{ij}(0)|$ if $i \ne j$.

Now, we consider multivariate white noise white series, which are analogous to that of univariate white noise. The m-variate series $\{z_t\}$ is called white noise with mean **0** and covariance matrix Σ as

$$\{z_t\} \sim WN(\mathbf{0}, \Sigma)$$

if $\{z_t\}$ is stationary with mean vector **0** and covariance matrix function

$$\Gamma(h) = \begin{cases} \sum & \text{if } h = 0 \\ 0 & \text{otherwise} \end{cases}$$

The m-variate series $\{z_t\}$ is called IID noise with mean **0** and covariance matrix Σ as

$$\{z_t\} \sim IID(\mathbf{0}, \Sigma)$$

provided the random vectors are $\{z_t\}$ are iid with mean **0** and covariance matrix Σ.

Multivariate white noise $\{z_t\}$ is used as a building block to construct a variety of multivariate time series.

The linear processes are generated as follows:

The m-variate series $\{X_t\}$ is a linear process if it has the representation

$$X_t = \sum_{j=-\infty}^{\infty} c_j z_{t-j}, \{z_t\} \sim WN(\mathbf{0}, \Sigma)$$

where $\{c_j\}$ is a sequence of $m \times m$ matrices whose components are absolutely summable. This linear process is stationary with mean $\mathbf{0}$ and covariance function

$$\Gamma(h) = \sum_{j=-\infty}^{\infty} c_{j+h} \sum c_j', h = 0, \pm 1, \pm \dots$$

An MA(∞) process is a linear process with $c_j = 0$ for $j < 0$.

Thus, $\{X_t\}$ is a linear process if there exists a white noise sequence $\{z_t\}$ and a sequence of matrices c_j with absolutely summable components so that

$$X_t = \sum_{j=0}^{\infty} c_j \, z_{t-j}$$

Any causal ARMA(p, q) process can be expressed as an MA(∞) process, while any invertible ARMA(p, q) process can be expressed as an AR(∞) process, i.e.

$$X_t + \sum_{j=1}^{\infty} A_j X_{t-j} = z_t$$

in which matrices A_j have absolutely summable components.

In the frequency domain, Γ has a matrix-valued spectral density function

$$f(\lambda) = \frac{1}{2\pi} \sum_{h=-\infty}^{\infty} e^{-i\lambda h} \, \Gamma(h), -\pi \le \lambda \le \pi$$

and Γ can be expressed in terms of f as:

$$\Gamma(h) = \int_{-\pi}^{\pi} e^{-i\lambda h} f(\lambda) d\lambda$$

provided components of covariance matrix function $\Gamma(\cdot)$ have property that

$$\sum_{h=-\infty}^{\infty} |\gamma_{ij}(h)| < \infty, i, j = 1, \dots, m.$$

The second order properties of stationary process $\{X_t\}$ can then be expressed in terms of $f(\cdot)$ rather than $\Gamma(\cdot)$.

$\{X_t\}$ has spectral representation $X_t = \int_{-\pi}^{\pi} e^{-i\lambda t} dz(\lambda)$

where $\{z(\lambda), -\pi \le \lambda \le \pi\}$ is a process whose components are complex-valued processes satisfying the following:

$$E(dz_j(\lambda)\, d\bar{z}_k(\mu)) = \begin{cases} f_{jk}(\lambda)d\lambda & \text{if } \lambda = \mu \\ 0 & \text{if } \lambda \ne \mu \end{cases}$$

and $\bar{z}_k$ is a complex conjugate of z_k.

We can show the approximately normal distribution of $\overline{X}_n$ for large n on the basis of central limit theorem also. Then the covariance matrix of $\{X_t\}$ and this will allow an estimation of confidence regions for μ. But this method is very complicated and a simpler method would be through the use of spectral density of ith process $\{X_{ti}\}$. Specifically, if n is large, then $\sqrt{n}(\overline{X}_i - \mu_i)$ is approximately normally distributed with mean zero and variance

$$2\pi f_i(0) = \sum_{k=-\infty}^{\infty} \gamma_{ii}(k)$$

We also know $2\pi \hat{f}_i(0) = \sum_{|h| \le r} \left(1 - \frac{|h|}{r}\right)\hat{\gamma}(h)$ is a consistent estimator of $2\pi f_i(0)$ provided $\gamma = \gamma_n$ is a sequence of numbers dependent on n in such a way that $\gamma_n \to \infty$ and $\gamma_n/n \to 0$ and $n \to \infty$.

Thus, if $\overline{X}_i$ denotes the sample mean of the ith process and Φ_α is the α-quantile of standard normal distribution, then the bounds $\overline{X}_i \pm \Phi_{1-\alpha/2}$ $(2\pi \hat{f}_i(0)/n)^{1/2}$ are asymptotic $(1 - \alpha)$ confidence bounds for μ_1. Therefore, as $n \to \infty$, the set of m-dimensional vectors bounded by $\{x_i = \overline{X}_i \pm \phi_{1-(\alpha/2m)} (2\pi \hat{f}_i(0)/n)^{1/2}, i = 1, 2, \ldots, m\}$ has a confidence coefficient which converges to a value greater than or equal to $(1 - \alpha)$ (and substantially greater especially if m is large).

We can estimate $\hat{\Gamma}(h)$ as:

$$\hat{\Gamma}(h) = \begin{cases} n^{-1}\sum_{t=1}^{n-h}(X_{t+h} - \overline{X}_n)(X_t - \overline{X}_n)', & \text{for } 0 \le h \le n-1 \\ \hat{\Gamma}(-h), & \text{for } -n+1 \le h < 0 \end{cases}$$

which gives $\hat{\rho}_{ij} = \hat{\gamma}_{ij}(h)/(\hat{\gamma}_{ii}(0)\hat{\gamma}_{jj}(0))^{1/2}$.

Let $\{X_t\}$ be bivariate time series whose components are

$$X_{t1} = \sum_{k=-\infty}^{\infty} \alpha_k z_{t-k,1} \, , \, \{z_{t1}\} \sim \text{IID}(0, \, \sigma_1^2)$$

$$X_{t2} = \sum_{k=-\infty}^{\infty} \beta_k z_{t-k,2} \, , \, \{z_{t2}\} \sim \text{IID}(0, \, \sigma_2^2)$$

where sequence $\{z_{t1}\}$ and $\{z_{t2}\}$ are independent, $\sum_k |\alpha_k| < \infty$ and $\sum_k |\beta_k| < \infty$.

Then, for all integers h and k with $h \neq k$, the random variables $n^{1/2} \hat{\rho}_{12}(h)$ and $n^{1/2} \hat{\rho}_{12}(k)$ are approximately bivariate normal with mean 0, variance $\sum_{j=-\infty}^{\infty} \rho_{11}(j)\rho_{22}(j)$ and covariance $\sum_{j=-\infty}^{\infty} \rho_{11}(j)\rho_{22}(j+-h)$ for large n.

If one of the processes is white noise, then $\hat{\rho}_{12}(h)$ is approximately normal with mean 0 and variance $1/n$ and H_0: $\rho_{12}(h) = 0$ and H_1: $\rho_{12}(h) \neq 0$ is straightforward.

But if both the series are independent but neither of them is white noise, then $\hat{\rho}_{12}(h)$ is not normally distributed and H_0: $\hat{\rho}_{12}(h)$ is not admissible for hypothesis testing.

Since large-sample distribution of $\hat{\rho}_{12}(h)$ depends on $\rho_{11}(\cdot)$ and $\rho_{22}(\cdot)$, any test of independence of the two component series cannot be based solely on the estimated values of $\rho_{12}(h)$, $h = 0, \pm 1, \pm \ldots$ without taking into account the nature of the two component series. This problem can be solved by 'prewhitening' the two series before computing the cross-correlations $\hat{\rho}_{12}(h)$, i.e. by transforming the two series to white noise by application of suitable filters. If $\{X_{t1}\}$ and $\{X_{t2}\}$ are invertible ARMA(p, q) processes, this can be achieved by the transformations

$$z_{ti} = \sum_{j=0}^{\infty} \pi_j^{(i)} X_{t-j,i}$$

where $\sum_{j=0}^{\infty} \pi_j^{(i)} B^j = \phi^{(i)}(B) / \theta^{(i)}(B)$.

Since the true model is unknown, it is convenient to replace the sequences $\{z_{ti}\}$ by the residuals $\{\hat{W}_{ti}\}$, using the estimated parameters of selected model (assumed to be true model which is a white noise sequence for $i = 1$ or 2).

Thus, under H_0, the corresponding two white noise series $\{z_{t1}\}$ and $\{z_{t2}\}$ are also independent. The sample autocorrelations $\hat{\rho}_{12}(h)$ and $\hat{\rho}_{12}(k)$, $h \neq k$ of $\{z_{t1}\}$ and $\{z_{t2}\}$ are for large n approximately independent and normally distributed with mean **0** and variance $n - 1$ and $\hat{\rho}_{12}(h)$ under H_0 also falls within $\pm 1.96 \sqrt{n}$ with a probability of approximately 0.95.

Example 1

If $\{X_{ti}\}$ are AR(1) processes for $i = 1, 2, \ldots$ with parameters $\phi_1 = \phi_2$ or $\phi_1 \neq \phi_2$, find the cross-correlation function $\hat{\rho}_{12}(h)$ and show that $\hat{\rho}_{12}(h) \neq 0$ in general.

Let $\rho_{11}(h) = \rho_{22}(h) = .8^{|h|}$ in the first case and $\rho_{11}(h) = .8^{|h|}$ and $\rho_{22}(h) = (-.8)^{|h|}$ in the second case.

Then large sample variances of $\hat{\rho}_{12}(h)$ in the first case are

$$n^{-1}\left(1 + 2\sum\nolimits_{k=1}^{\infty}(.8)^{2k}\right) = 4.55\, n^{-1}.$$

In the second case, $\hat{\rho}_{12}(h) = 0.2195\, n^{-1}$. An observed value of $\hat{\rho}_{12}(h)$ as large as $3n^{-1/2}$ would be likely even though $\{X_{t1}\}$ and $\{X_{t2}\}$ are independent, for the first case.

For the second case, an observed value of $\hat{\rho}_{12}(h)$ could be very unlikely to be $3n^{-1/2}$.

This shows that cross-correlations between time series is useless for independence tests unless at least one of the series is a white noise.

If $\{X_t\}$ is a bivariate Gaussian time series with covariances satisfying

$$\sum\nolimits_{-\infty}^{\infty}|\gamma_{ij}(h)| < \infty, i, j = 1, 2, \ldots$$

then

$$\lim_{n\to\infty} n\, Cov\,(\hat{\rho}_{12}(h), \hat{\rho}_{12}(k))$$

$$= \sum\nolimits_{j=-\infty}^{\infty}\begin{bmatrix} \rho_{11}(j)\rho_{22}(j+k-h) + \rho_{12}(j+k)\rho_{2}(j-h) \\ -\rho_{12}(h)\{\rho_{11}(j)\rho_{12}(j+k) + \rho_{22}(j)\rho_{21}(j-k)\} \\ -\rho_{12}(k)\{\rho_{11}(j)\rho_{12}(j+h) + \rho_{12}(j)\rho_{21}(j-h)\} \\ +\rho_{12}(h)\rho_{12}(k)\left\{\frac{1}{2}\rho_{11}^2(j) + \rho_{12}^2(j) + \frac{1}{2}\rho_{22}^2(j)\right\} \end{bmatrix}$$

If $\{X_t\}$ satisfies the conditions for the above formula, either $\{X_{t1}\}$ or $\{X_{t2}\}$ is white noise and if $\rho_{12}(h) = 0$, $h \notin [a, b]$, then

$$\lim_{n \to \infty} nVar(\hat{\rho}_{12}(h)) = 1, h \notin [a, b].$$

Example 2

Sales series and its leading indicator series were observed with 150 data for each series. The first differenced series for each were found to be stationary and ARMA models gave

$$\nabla X_{t1} - .0228 = z_{t1} - .474\, z_{t-1,1},\ \{z_{t1}\} \sim WN(0,0.779)$$

$$\nabla X_{t2} - .838\nabla X_{t-1,2} - .0676 = z_{t2} - .610\, z_{t-1,2},\ \{z_{t2}\} \sim WN(0,1.754)$$

The sample cross-correlation ∇X_{t1} and ∇X_{t2} is meaningless, since none of them is white noise. However, sample cross-correlation between whitened series $\{\hat{W}_{t1}\}$ and $\{\hat{W}_{t2}\}$ could be more informative regarding the independence and the correlation structure of the two series $\{X_{t1}\}$ and $\{X_{t2}\}$.

We find that $\rho_{12}(-3) \neq 0$ and $\rho_{12}(h) = 0$, $h \neq -3$.
The value of $\rho_{12}(-3) = .969$ suggests strong dependence at $h = -3$ and the model is

$$\hat{W}_{t2} = 4.74\, \hat{W}_{t-2,1} + N_t$$

where $\{N_t\}$ has small variance compared with those of $\{\hat{W}_{t2}\}$ and $\{\hat{W}_{t1}\}$.

We model sample values of $\{\hat{W}_{t2} - 4.74\, \hat{W}_{t-3}\}$ for $\{N_t\}$ to obtain

$$(1 + .345B)\, N_t = U_t,\ \{U_t\} \sim WN(0,.0782)$$

So, the model in terms of ∇X_{t1} and ∇X_{t2} is

$$\nabla X_{t,2} + .0773 = (1 - 0.610B)\,(1 - .838B)^{-1}\, \nabla X_{t,2}\, [4.74(1 - .474B)^{-1}\nabla X_{t-3,1} + (1 + .345B)^{-1}U_t].$$

A better approach could be to use Transfer Function Modelling.

4.2 MULTIVARIATE ARMA PROCESSES

$\{X_t\}$ is an ARMA(p, q) process if $\{X_t\}$ stationary and if for any t,

$$X_t - \Phi_1 X_{t-1} - \cdots - \Phi_p X_{t-p} = z_t + \Theta_1 z_{t-1} + \cdots + \Theta_q z_{t-q}$$

where $\{z_n\} \sim WN(0, \Sigma)$, $\{X_t\}$ is an ARMA(p, q) process with mean μ, if $\{X_{t-\mu}\}$ is an ARMA(p, q) process.

It can be completely written as:

$$\Phi_p(B)X_t \ = \ \Theta_q(B)z_t, \ \{z_t\} \sim \mathrm{WN}(0,\, \Sigma)$$

Example 3

Suppose $p = 1$ and $q = 0$ which gives

$$X_t \ = \Phi X_{t-1} + z_t, \ \{z_t\} \sim \mathrm{WN}(0,\, \Sigma)$$

for multivariate AR(1) process.
We can expand X_t as

$$X_t \ = \sum_{j=0}^{\infty} \Phi^j z_{t-j}$$

provided all roots are less than 1 in absolute values, i.e. det $(I - \Phi_z) \neq 0$ for all z in C such that $|z| \leq 1$.

This is satisfied when ϕ^j are absolutely summable and hence the series converges, i.e. each component of matrix $\sum_{j=0}^{\infty} \Phi^j z_{t-j}$ converges.

Causality and invertibility conditions are same as for univariate time series, with Φ replacing ψ_j in MA(∞) model and π_j matrices for AR(∞) representations with components that are absolutely summable.

An ARMA(p, q) of $\{X_t\}$ is the causal or causal function of $\{z_t\}$ if there exist matrices $\{\psi_j\}$ with absolute summable components such that $X_t = \sum_{j=0}^{\infty} \psi_j z_{t-j}$ for all t, which is equivalent to det $(\Phi_z) \neq 0$ for all $z \in$ C such that $|z| \leq 1$.

Matrices ψ_j are found recursively for $\psi_j = \Theta_j + \sum_{k=1}^{\infty} \phi_k \psi_{j-k}$, $j = 0, 1, \ldots$

where $\Theta_0 = I$, $\Theta_j = 0$ for all $j > q$; $\phi_j = 0$ for $j > p$; $\psi_j = 0$ for $j < 0$.
An ARMA(p, q) $\{X_t\}$ is invertible if there exist matrices $\{\pi_j\}$ with absolutely summable components such that

$$z_t \ = \sum_{j=0}^{\infty} \pi_j X_{t-j} \ \text{for all } t.$$

This is equivalent to condition det $(\Phi_z) \neq 0$ for all $z \in$ C such that $|z| \leq 1$.
The matrices π_j are found recursively for the equations

$$\pi_j \ = -\Phi_j - \sum_{k=1}^{\infty} \Theta_k \, \pi_{j-k}, \, j = 0, 1, \ldots$$

where we define $\Phi_0 = -I$, $\Phi_0 = 0$ for $j > p$, $\Theta_j = 0$ for $j > q$ and $\pi_j = 0$ for $j < 0$.

Example 4

Multivariate AR(1) is given by

$$\psi_0 = I, \ \psi_1 = \Phi\psi_0 = \Phi, \ \psi_2 = \Phi\psi_1 = \Phi_2, \ ...., \ \psi_j = \Phi\psi_{j-1} = \Phi\psi_j, \ j \leq 3.$$

For bivariate AR(1) process with $\Phi = \begin{bmatrix} 0 & 1/2 \\ 0 & 0 \end{bmatrix}$, we check that $\psi_j = \Phi^j = 0$

for $j > 1$ and hence $\{X_t\}$ has the alternative representation

$$X_t = z_t + \Phi z_{t-1}$$

as an MA(1) process. So, we cannot distinguish between multivariate ARMA models of different orders without imposing further restrictions. If we always model multivariate time series as pure AR processes, then the problem vanishes.

The covariance matrix function of a causal ARMA process is given by

$$\Gamma(h) = E(X_{t+h}X'_t)$$

and in terms of model parameters

$$\Gamma(h) = \sum_{j=0}^{\infty} \psi_{h+j}\Sigma\psi'_j, \ h = 0, \pm 1, ...$$

where $\psi_j = 0$ for $j < 0$.

The covariance matrices $\Gamma(h)$, $h = 0, \pm 1, ...$ can also be found by solving

this Yule-Walker equations $\Gamma(j) - \sum_{r=1}^{p} \phi_r \Gamma(j - r) = \sum_{j \leq r \leq q} \Theta_r \Sigma\psi_{r-j}$ obtai-

ned by post-multiplying X'_{t-j} and taking expectations.

The first $p + 1$ equations are solved for $\Gamma(0)$ to $\Gamma(p)$ with the known fact that $\Gamma(h) = \Gamma(-h)$. The remaining equations then give $\Gamma(p + 1), \Gamma(p + 2)$, ... recursively.

If z_0 is the root of $\det \Phi(z) = 0$ with the smallest absolute value, then it is easy to show that $\psi_j/r^j \rightarrow 0$ as $j \rightarrow \infty$ for all r such that $|z_0|^{-1} < r < 1$. Hence, the constant C exists in such a manner that each component of ψ_j is smaller in absolute value than Cr^j. This implies that there exists a constant κ so that each component of matrix $\psi_{n+j}\Sigma\psi'_j$ is bounded in absolute value by κ^{2j}. Provided that $|z_0|$ is not very close to 1, this means that $\Gamma(h)$ rapidly converges and the error increased in each component by truncating the series after terms with $j = (\kappa - 1)$ is a smaller absolute value

than $\sum_{j=k}^{\infty} \kappa r^{2j} = \kappa r^{2k} / (1 - r^2)$.

4.2.1 Best Linear Predictors of Second-order Random Vectors

Let $\{X_t = (X_{t1}, ..., X_{tm})'\}$ be an m-variate time series with means $EX_t = \mu_t$ and covariance function given by matrices

$$K(i, j) = E(X_i, X_j') - \mu_i \mu_j'$$

If $Y = (Y_1, ..., Y_m)'$ is a random vector with finite second moments and $EY = \mu$, we define

$$P_n(Y) = (P_n Y_1, ..., P_n Y_m)'$$

where $P_n Y_j$ is the best linear predictor of the component Y_j of Y in terms of all the components of vectors X_t, $t = 1, 2 ..., n$ and the constant 1. It follows immediately from the properties of prediction operator that

$$P_n(Y) = \mu + A_1(X_n - \mu_n) + \cdots + A_n(X_1 - \mu_1)$$

For some matrices $A_1, ..., A_n$ and that

$$Y - P_n(Y) \perp X_{n+1-i}, i = 1, ..., n.$$

where we say two m-dimensional vectors X and Y are orthogonal if $E(XY')$ is a matrix of zeros. $P_n(Y)$ is unique, although there may be more than one possible choice for $A_1, ..., A_n$.

As a special case, we consider $\{X_t\}$ to be a zero mean time series and X_{n+1} is the best linear predictor of X_{n+1} in terms of $X_1 ... X_n$. This is solved by replacing Y with X_{n+1} and we get

$$\hat{X}_{n+1} = \begin{cases} 0 & \text{if } n = 0 \\ \rho_r(X_{n+1}) & \text{if } n \geq 1 \end{cases}$$

Hence, we can write

$$\hat{X}_{n+1} = \Phi_{n1} X_n + \cdots + \Phi_{nn} X_1, n = 1, 2, ...$$

where coefficients $\Phi_{nj}, j = 1, 2, ... n$ are such that

$$\sum_{j=1}^{n} \Phi_{nj} k(n + 1 - j, n + 1 - i) = k(n + 1, n + 1 - i), i = 1, ..., n$$

In the case when $\{X_t\}$ is stationary with $k(i, j) = \Gamma(i - j)$, the prediction equations simplify to the m-dimensional form as

$$\sum_{j=1}^{n} \phi_{nj} \Gamma(i - j) = \Gamma(i), i = 1, 2 ... n$$

Provided the covariance matrix of the nm components of $X_1 \dots X_n$ are non-singular for every $n \geq 1$, the coefficients $\{\Phi_{nj}\}$ can be determined recursively, using a multivariate version of the Durbin-Levison algorithm.

This also determines the covariance matrices of the one-step prediction errors $V_0 = \Gamma(0)$, and for $n \geq 1$,

$$V_n = E\,(X_{n+1} - \hat{X}_{n+1})\,(X_{n+1} - \hat{X}_{n+1})'$$

$$= \Gamma(0) - \phi_{n1}\Gamma(-1) - \cdots - \phi_{nn}\Gamma(-n)$$

We can also use innovations algorithm with a multivariate version (instead of the Durbin-Levison algorithm) for prediction purposes.

4.2.2 Modelling Multivariate AR Process

If $\{X_t\}$ is any zero-mean second order multivariate time series, it is easy to show from results of the previous section that one-step prediction errors $X_j - \hat{X}_j$, $j = 1, \dots, n$ have the property

$$E(X_j - \hat{X}_j)\,(X_k - \hat{X}_k)' = 0 \text{ for } j \neq k.$$

Moreover, matrix M is such that

$$\begin{bmatrix} X_1 & \hat{X}_1 \\ X_2 & \hat{X}_2 \\ \dots & \dots \\ X_n & \hat{X}_n \end{bmatrix} = M \begin{bmatrix} X_1 \\ X_1 \\ \dots \\ X_1 \end{bmatrix}$$

is linear triangular with ones on the diagonal and, therefore, has determinant equal to 1.

If the series $\{X_t\}$ is also Gaussian, then the prediction errors $U_j = X_j - \hat{X}_j$, $j = 1, \dots, n$, are independent with covariance matrices $V_0, \dots, V_{n-1}$ recursively. Consequently, the joint density of prediction errors is the product

$$f(u_1, \dots, u_n) = (2\pi)^{-nm/2} \left(\prod_{j=1}^{n} \det V_{j-1} \right)^{1/2} \exp\left[\frac{1}{2}\sum_{j=1}^{n} u_j'\,V_{j-1}^{-1}\,u_j \right]$$

We can replace u_j by $X_j - \hat{X}_j$ in the last expression corresponding to observations $x_1, \ldots, x_n$ since determinant M is known to be 1.

If we suppose that $\{X_t\}$ is a zero-mean m-variate AR(p) process with coefficient matrices $\Phi = \{\Phi_1, \ldots \Phi_p\}$ and white noise covariance matrix Σ, we can express likelihood of observations $X_1 \ldots X_n$ as

$$L(\Phi, \Sigma) = (2\pi)^{-nm/2} \left(\prod_{j=1}^{n} \det V_{j-1} \right)^{1/2} \exp\left[\frac{1}{2} \sum_{j=1}^{n} u_j' V_{j-1}^{-1} u_j \right]$$

where $U_j = X_j - \hat{X}_j$, $j = 1, \ldots, n$ and V_j was as given above.

Maximization of the Gaussian likelihood is much more difficult in the multivariate case than in the univariate case because of the potentially large number of parameters involved and we cannot compute MLE of Φ independently of Σ as in the univariate case. For pure AR processes, good preliminary estimates for later linear optimization algorithms can be obtained by either Whittle's algorithm in multivariate version of Burg's algorithm given by Jones (1978). Other frequency domain methods can also be used. The order selection for multivariate AR models can be made by minimizing a multivariate that is analogous of univariate AICC statistics as

$$\text{AICC} = -2 \ln L(\Phi_1, \ldots \Phi_p, \Sigma) + \frac{2(pm^2 + 1)nm}{nm - pm^2 - 2}$$

We now use the Whittle's algorithm to estimate the parameters for a causal multivariate AR(p) process defined by the difference equations

$$X_t = \phi_1 X_{t-1} + \cdots + X_{t-p} + z_t; \ \{z_t\} \sim WN(0, \Sigma)$$

Then, post-multiplying by X_{t-j}', $j = 0, \ldots p$ and taking expectations gives the equations:

$$\Sigma = \Gamma(0) - \sum_{j=1}^{p} \Phi_j \Gamma(-j)$$

and

$$\Gamma(i) = \sum_{j=1}^{n} \Phi_j \Gamma(i - j), i = 1, \ldots, p$$

Given the matrices $\Gamma(0), \ldots, \Gamma(p)$, we can solve above equations to obtain $\phi_1, \ldots \phi_p$.

In this case, the white noise covariance matrix Σ can be easily found. These solutions are the same as those of solution to prediction of error covariance matrix V_p. Thus, Whittle's algorithm is correct.

4.2.3   Forecasting Multivariate AR Processes

We can obtain minimum MSE one-step linear predictors $\hat{X}_{n+1}$ for any multivariate stationary time series for the autocovariance matrices $\Gamma(h)$ by recursively determining the coefficients ϕ_{ni}, $i = 1, \ldots n$ and evaluating

$$\hat{X}_{n+1} = \phi_{n1}X_n + \cdots + \phi_{nn}X_1$$

The situation is simplified when $\{X_t\}$ is an AR(p) process since for $n \geq p$, which is usually the case

$$\hat{X}_{n+1} = \phi X_n + \cdots + \phi_p X_{n+1-p}$$

since $X_{n+1} - \hat{X}_{n+1} = z_{n+1}$ which is orthogonal to $X_1, \ldots, X_n$.

The covariance matrix of the one-step prediction error is clearly $E(z_{n+1}, z'_{n+1}) = \Sigma$.

To compute the best h-step linear predictor $P_n X_{n+h}$ based on all the components of $X_1, \ldots, X_n$, we apply the linear operator P_n to the AR difference equations in order to obtain the recursive

$$P_n X_{n+h} = \phi_1 P_n X_{n+h-1} + \cdots + \phi_p P_n X_{n+h-p}$$

These equations are easily solved, first for $P_n X_{n+1}$, then for $P_n X_{n+2}$, ... etc.

If $n \geq p$, then the h-step predictions based on all components of X_j, $-\infty < j \leq n$ also satisfy the linear predictor equation based on P_n and are same as the h-step predictors based on $X_1, \ldots, X_n$.

To compute h-step error covariance matrices, we have

$$X_{n+h} = \sum_{j=0}^{\infty} \psi_j z_{n+h-j}$$

where ψ_j are obtained for ARMA(p, q) causal process with $q = 0$. We, therefore, obtain for

$$P_n X_{n+h} = \sum_{j=h}^{\infty} \psi_j z_{n+h-j}$$

Therefore, $X_{n+h} - P_n X_{n+h} = \sum_{j=0}^{h-1} \psi_j z_{n+h-j}$ gives the h-step predictor error for causal AR(p) process and its covariance matrix is given by

$$E[(X_{n+h} - P_n X_{n+h})(X_{n+h} - P_n X_{n+h})'] = \sum_{j=0}^{h-1} \psi_j \Sigma \psi_j' , n \geq p.$$

The above calculations are based on the assumption that the AR(p) model for the observed time series is known. In practice, the parameters of the model are usually estimated for the data and hence the predicted values will be more uncertain than indicated above.

4.2.4 Cointegration Technique

It is somewhat restrictive to obtain stationarity for multivariate time series using the same differencing operator $D(B)$ for all the components of the vectors. We can use more general linear transformations for each component of the vector to produce stationarity for modelling purposes which is termed cointegration.

If $\nabla^d X_t$ is stationary but $\nabla^{d-1} X_t$ is non-stationary for $d > 0$, then we say $\{X_t\}$ is integrated of order d, i.e. $\{X_t\} \sim I(d)$. Many economic time series are often integrated of order 1.

If $\{X_t\}$ is a k-variate time series, we define $\{\nabla^{d-1} X_t\}$ to be the series where jth component is obtained by applying operator $(1 - B)^d$ to the jth component of $\{X_t\}$, $j = 1, ..., k$.

We define k-dimensional time series $\{X_t\}$ is integrated of order d (or $\{X_t\} \sim I(d)$) if d is a positive integer, $\{\nabla^d X_t\}$ is stationary but $\{\nabla^{d-1} X_t\}$ is non-stationary.

The $I(d)$ process $\{X_t\}$ is said to be cointegrated with cointegration vector α if α is a $k \times 1$ vector such that $\{\alpha' X_t\}$ is of order less than d.

A simple example is the bivariate process whose first component is the random walk

$$X_t = \sum_{j=1}^{t} z_j, t = 1, 2 ..., \{z_t\} \sim \text{IID}(0, \sigma^2)$$

and whose second noisy observations of the same random walk, i.e.

$$Y_t = X_t + W_t, t = 1, 2, ..., \{W_t\} \sim WN(0, \zeta^2)$$

where $\{W_t\}$ is independent of $\{z_t\}$. Then, $\{(X_t, Y_t)'\}$ is integrated of order 1 and cointegrated with cointegration vector $\alpha = (1, -1)'$.

The concept of cointegration subsumes the idea of univariate non-stationary time series 'moving together'. Thus, $\{X_t\}$ and $\{Y_t\}$ are both non-stationary but they are linked to move together as they differ only in the stationary sequence $\{W_t\}$.

As an example, prices of oil in the USA and Middle East countries may behave as cointegrated series as if Middle East oil price reduces sufficiently below the US price, US production may be reduced and oil imported from Middle East and sold at US localities. This pushes the non-stationary prices u_t and v_t at these two to be stabilized towards a straight line attractor $v = u$ in R^2. The points $(u_t, v_t)'$ will exhibit small random deviations for the line $v = u$.

Example 5

We apply $\nabla = (1 - B)$ operator to the bivariate series defined above in order to obtain $(u_t, v_t)'$, where $u_t = z_t$ and $v_t = z_t + W_t - W_{t-1}$.

The series $\{(u_t, v_t)'\}$ is clearly a stationary multivariate MA(1) process:

$$\begin{bmatrix} u_t \\ v_t \end{bmatrix} = \begin{bmatrix} 1 & 0 \\ 0 & 1 \end{bmatrix} \begin{bmatrix} z_t \\ z_t + W_t \end{bmatrix} - \begin{bmatrix} 0 & 0 \\ -1 & 1 \end{bmatrix} \begin{bmatrix} z_{t-1} \\ z_{t-1} + W_{t-1} \end{bmatrix}$$

But the process $\{[u_t, v_t]'\}$ cannot be represented as an AR(∞) process since the matrix

$$\begin{bmatrix} 1 & 0 \\ 0 & 1 \end{bmatrix} - z \begin{bmatrix} 0 & 0 \\ 1 & 1 \end{bmatrix}$$

has zero determinant when $z = 1$, thus violating condition $\det \Theta(z) \neq 0$.

Therefore, we must be very careful in estimating the parameters of such cointegration models assuming these as error-correction models.

4.3   TRANSFER FUNCTION MODELS

4.3.1   Introduction

Transfer function models are used to exploit for forecasting purposes the relationship between when one acts as a leading indicator for the other. We wish to estimate the transfer function of a linear filter when output includes added uncorrelated noise. Suppose $\{X_{t1}\}$ and $\{X_{t2}\}$ are, respectively, input and output series for the transfer function model, i.e.

$$X_{t2} = \sum_{j=0}^{\infty} \tau_j X_{t-j,1} + N_t$$

where $T = \{\tau_j, j = 0, 1 \ldots\}$ is a causal time-invariant linear filter and $\{N_t\}$ is a zero-mean stationary process uncorrelated with input $\{X_{t1}\}$. We also assume that $\{X_{t1}\}$ is a zero-mean stationary series. Then the bivariate process $\{(X_{t1}, X_{t1})'\}$ is also stationary.

Multiplying both sides by $\{X_{t-k}, 1\}$ and taking expectations gives us

$$\gamma_{21}(k) = \sum_{j=0}^{\infty} \tau_j \gamma_{11}(k-j)$$

If input $\{X_{t1}\}$ is WN$(0, \sigma_1^2)$, then, we get

$$\tau_k = \gamma_{21}(k)/\sigma_1^2$$

This clearly suggests that 'pre-whitening' of the input(s) might simplify this identification of appropriate transfer function model and, at the same time, provide simple preliminary estimates of the coefficients, t_k.

If $\{X_{t1}\}$ is causal invertible ARMA(p, q)

$$\phi_p(B)X_{t1} = \theta_q(B)z_t, \ \{z_t\} \sim WN(0, \sigma_z^2)$$

then the application of filter $\pi(B) = \phi(B)\theta^{-1}(B)$ to $\{X_{t1}\}$ will produce the whitened series $\{z_t\}$.

Now applying the operator $\pi(B)$ to each side of equation above and letting $Y_t = \pi(B)X_{t2}$, we obtain

$$Y_t = \sum^{\infty} \tau_j z_{t-j} + N_t'$$

where $N_t' = \pi(B)N_t$ and $\{N_t'\}$ is a zero-mean stationary process uncorrelated with $\{z_t\}$.

So, we obtain

$$\tau_j = \rho_{Yz}(j)\sigma_Y/\sigma_z$$

where σ_{Yz} is the cross-correlation function of $\{Y_t\}$ and $\{z_t\}$, $\sigma_z^2 = \text{Var}(z_t)$ and $\sigma_Y^2 = \text{Var}(Y_t)$.

The steps described in next section are to be strictly followed for estimating $\{\tau_j\}$ and analyzing the noise $\{N_t\}$ in the above transfer function model (see Fig. 4.2).

Example 5 *Multi-dimensional Series*

In earth sciences, we often have map data recorded on the surface of the earth. These data may either have been collected in a rectangular/square grid or the map may be digitized into a rectangular/square grid data pattern, which seems to be a rather stringent requirement. In remote sensing, often the pixel data are arranged in a rectangular/square grid. These data become easily amenable to the multi-dimensional method. A vector (column) time series $x_t = (x_{t1}, x_{t2}, ..., x_{tp})'$ contains p univariate time series. For stationary cases, we have $\mu = E(x_t)$ and $p \times p$ autocovaraince matrix

$$\Gamma(h) = E[(x_t - h - \mu)(x_t - \mu)']$$

where elements of $\Gamma(h)$ are the cross-covariance function functions

$$\gamma_{ij}(h) = E[(x_{t+h,i} - \mu_i)(x_{tj} - \mu_j)]$$

Since $\gamma_{ij}(h) = \gamma_{ij}(-h)$, it follows that $\Gamma(-h) = \Gamma'(h)$.

For sample data, we replace population values by corresponding sample statistics $\bar{x}$ and $\hat{\Gamma}(h)$.

For multi-dimensional process, we have x_s as a function of $r \times 1$ vectors $s = (s_1, s_2, ..., s_r)'$, where s_i denotes coordinates of the ith index.

We give a two-dimensional variation of soil temperature series x_{s1}, x_{s2} indexed by row numbers s_1 and column numbers s_2, representing positions on a 64×36 grid (with spacing of 17 ft) in an agricultural field.

Here $\qquad \mu = E(x_s), \; \gamma(h) = E[(x_{s+h} - \mu)(x_s - \mu)]$

and $\qquad \gamma(h_1, h_2) = E[(X_{s_1+h_1,s_2+h_2} - \mu)(x_{s_1,s_2} - \mu)]$

where h_1 and h_2 are lags for rows and columns, respectively.

The multi-dimensional sample autocorrelations are $\hat{\rho}(h) = \hat{\gamma}(h)/\gamma(0)$.

The row temperatures are cyclically oscillating greatly for 0 to 45 and thereafter very little for 45 to 64, whereas along columns, they show random fluctuations about mean levels. The row arrays of temperature values seem cyclic with period of 17 rows (i.e. $17 \times 17 = 289$ ft).

The two-dimensional autocorrelation function $\hat{\rho}(h_1, h_2) = \hat{\gamma}(h_1, h_2)/\gamma(0, 0)$ shows a sharp peak at (15, 10) and a quick decrease along the rows and columns (Fig. 4.1).

In the frequency domain, the multi-dimensional wave number (frequency) spectrum is given as the Fourier transform of autocovariance as

$$f_x(v) = \sum_h \gamma_x(h) e^{-2\pi i v'h}$$

which, on inversion, gives

$$\gamma_x(h) = \int_{-1/2}^{1/2} f_x(v) e^{2\pi i v'h} \, dv$$

The cycling rate is v_i per the distance travelled s_i along the ith direction.

For a two-dimensional process, we get

$$f_x(v_1, v_2) = \sum_{h_1=-\infty}^{\infty} \sum_{h_2=-\infty}^{\infty} \gamma_x(h_1, h_2) e^{-2\pi i(v_1 h_1 + v_2 h_2)}$$

and $\quad \gamma_x(h_1, h_2) = \int_{-1/2}^{1/2} \int_{-1/2}^{1/2} f_x(v_1, v_2) e^{2\pi i(v_1 h_1 + v_2 h_2)} dv_1 dv_2$

The concept of linear filtering in 2D generalizes to by defining the impulse ratio function a_{s_1, s_2} and the spatial filter output as

$$y_{s_1, s_2} = \sum_{u_1} \sum_{u_2} a_{u_1, u_2} x_{s_1 - u_1, s_2 - u_2}$$

Thus, spectra of output of this filter is

$$f_y(v_1, v_2) = |A(v_1, v_2)|^2 f_u(v_1, v_2)$$

where $\quad A(v_1, v_2) = \sum_{u_1} \sum_{u_2} a_{u_1, u_2} e^{-2\pi i(v_1 u_1 + v_2 u_2)}$

The 2D Discrete Fourier transform (DFT) is given by

$$X(v_1, v_1) = (n_1 n_2)^{-1/2} \sum_{s_1=1}^{n_1} \sum_{s_2}^{n_2} x_{s_1, s_2} e^{-2\pi i(v_1 s_1 + v_2 s_2)}$$

where frequencies v_1, v_2 are evaluated at multiples of $(1/n_1, 1/n_2)$ on the spectral frequency scale. The 2D wavenumber spectrum is

$$\hat{f}_x(v_1, v_2) = (L_1 L_2)^{-1} \sum_{l_1 l_2} |X(v_1 + l_1/n_1, v_2 + l_2/n_2)|^2$$

where the sum is taken over the grid $\{-(L_j - 1)/2 \le j \le (L_j - 1)/2, j = 1, 2\}$, L_1, L_2 are odd.

The statistic is $\dfrac{2 L_1 L_2 \hat{f}_x(v_1, v_2)}{f_x(v_1, v_2)} \sim X_{2 L_1 L_2}^2$.

This is useful for setting confidence intervals or for tests of hypothesis against a fixed (assumed) spectrum $f_0(v_1, v_2)$.

The periodogram of 2D soil temperature data shows strong spectral peaks at frequencies 0.0625 and $-.0625$ cycles per row, corresponding to 16 rows ($16 \times 17 = 272$ ft), which is very similar to visual cycles observed on mean temperatures along different rows. This is due to salts which has put the fields periodically over columns for irrigation purposes (salt induces lower soil temperatures) (Fig. 4.1).

4.3.2 Steps for Transfer Function Modelling

(i) Fit an ARMA model to $\{X_{t1}\}$ and file the residuals $(\hat{z}_1, ..., \hat{z}_n)$.

Let $\hat{\phi}$ and $\hat{\theta}$ denote the MLE of the AR and MA parameters and σ_z^2 be the MLE of variance of $\{z_t\}$.

(ii) Apply the operator $\hat{\pi}(B) = \hat{\phi}(B)\hat{\theta}^{-1}(B)$ to $\{X_{t2}\}$ giving series $(\hat{Y}_1, ..., \hat{Y}_n)$.

Let σ_Y^2 denote the sample variance of $\hat{Y}_t$ (the "white noise" variance).

(iii) Compute cross-correlation function $\rho_{Yz}(h)$ between $\{\hat{Y}_t\}$ and $\{\hat{z}_t\}$.

A comparison of $\hat{\rho}_{Yz}(h)$ with bounds of $\pm 1.96 n^{-1/2}$ gives a prelimi-nary indication of the lags h at which is $\rho_{Yz}(h)$ significantly different for zero. We know that for transfer function, the model should be zero for $h < 0$.

(iv) Preliminary estimates of τ_h for lags h at which $\rho_{Yz}(h)$ is found to be significantly different for zero are

$$\hat{\tau}_h = \hat{\rho}_{Yz}(h)\hat{\sigma}_Y / \hat{\sigma}_z.$$

For other values of h, the preliminary estimates are $\hat{\tau}_h = 0$.

Let $m \geq 0$ be the largest value of j such that $\hat{\tau}_j$ is non-zero and let $b \geq 0$ be the smallest such value. Then b is known as the delay parameter of the filter $\{\hat{\tau}_j\}$.

If m is very large and if the coefficients $\{\hat{\tau}_j\}$ are approximately related by difference equations of the form

$$\hat{\tau}_j - v_1\hat{\tau}_{j-1} - \cdots - v_p\hat{\tau}_{j-p} = 0, j \geq b + p,$$

then $\hat{T}(B) = \sum_{j=h}^{m} \hat{\tau}_j B^j$ can be represented approximately using fewer parameters as

$$\hat{T}(B) = w_0(1 - v_1 B - \cdots - v_p B_p)^{-1} B^b$$

In particular, if $\hat{\tau}_j = 0, j < b$ and $\hat{\tau}_j = w_0 v_1^{j-b} \geq b$, then

$$\hat{T}(B) = w_0(1 - v_1 B)^{-1} B^b$$

Box and Jenkins (1976) suggested $\hat{T}(B)$ be ratio of two polynomials, but the degree of these polynomials is often difficult to estimate for $\{\hat{\tau}_j\}$.

If $\hat{T}(B) = B^b(w_0 + w_1 B + \cdots + w_q B^q)\,(1 - v_1 B + \cdots + v_p B^p)^{-1}$ with $v(z) \neq 0$ for $|z| \leq 1$, then we define $m = \max(q + b, p)$.

(v) The noise sequence $\{N_t, t = m + 1, ..., n\}$ is estimated by

$$\hat{N}_t = X_{t2} - \hat{T}(B)X_{t1}$$

We set $\hat{N}_t = 0, t \leq m$, in order that $\hat{N}_t, t > m = \max(b + q, p)$ can be computed.

(vi) Preliminary identification of a suitable model for a noise sequence is carried out by filtering a causal invertible ARMA model

$$\phi^{(N)}(B)N_t = \theta^{(N)}(B)W_t, \{W_t\} \sim WN(0, \sigma_N^2)$$

to the estimated noise $\hat{N}_{m+1}, ... \hat{N}_n$

(vii) Selection of parameters b, p and q and the orders p_2 and q_2 of $\phi^{(N)}(\cdot)$ and $\theta^{(N)}(\cdot)$ gives the preliminary model

$$\phi^{(N)}(B)v(B)X_{t2} = B^b w\,\phi^{(N)}(B)w(B)X_{t1} + \theta^{(N)}(B)v(B)W_t$$

where $\hat{T}(B) = B^b w(B)v^{-1}(B)$ as in step (iv).

We now compute $\hat{W}_t(w, v, \phi^{(N)}, \theta^{(N)})$, $t = m^*\max(p_2 + p, b + p_2 + q)$, by setting $\hat{W}_t = 0$ for $t < m^*$.

The parameters w, v, $\phi^{(N)}$, $\theta^{(N)}$ are estimated by minimizing

$$\sum_{t=m^*+1}^{n} \hat{W}_t^2(w, v, \phi^{(N)}, \theta^{(N)})$$

These preliminary values can be used as initial values in the minimization $\sum \hat{W}_t^2$ errors in transfer function modelling. Alternatively, MLE may be carried out using state-space representation (described in later section).

(viii) From the least square estimates of parameters of $T(B)$, a new estimated noise sequence can now be computed as in step (v) and checked for compatibility with ARMA model for $\{N_t\}$ fitted by LS procedure. If the new estimated noise sequence suggests different orders for $\phi^{(N)}(\cdot)$ and $\theta^{(N)}(\cdot)$, the LS procedure in step (vii) can be repeated using these new orders.

(ix) We can test goodness of fit, the residuals for ARMA fitting in steps (i) and (vi) should both be checked for normality and independence. The sample cross-correlations of the two residual series $\{\hat{z}_t, t > m^*\}$ and $\{\hat{W}_t, t > m^*\}$ should also be compared with bounds $\pm 1.96 n^{-1/2}$ to check whether $\{N_t\}$ and $\{z_t\}$ are uncorrelated.

4.3.3 Forecasting by Transfer Function Model

When predicting $X_{n+h,2}$ on the basis of transfer function model with observations of X_{t1} and, X_{t2}, $t = 1, 2, \ldots, n$ we wish to make linear forecasting on the basis of $1, X_{11}, X_{n1}, X_{11}, \ldots X_{n2}$ that predicts $X_{n+h,2}$ with minimum MSE. The exact solution to this problem can be found using Kalman recursions. We have predictors $P_n X_{n+h}$ and MSE based on infinite past observations X_{t1} and X_{t2}, $-\infty < t \le n$. These predictors and their MSE will be close to those based on $X_{t1}, X_{t2}, 1 \le t \le n$ provided n is sufficiently large.

The transfer function model can be written as

$$X_{t2} = T(B)X_{t1} + \beta(B)W_t$$

$$X_{t1} = \theta(B)\phi^{-1}(B)z_t.$$

where $\beta(B) = \theta^{(N)}(B)/\phi^{(N)}(B)$.
Eliminating X_{t1} gives

$$X_{t2} = \sum_{j=0}^{\infty} \alpha_j z_{t-j} + \sum_{j=0}^{\infty} \beta_j W_{t-j}$$

where $\alpha(B) = T(B)\theta(B)/\phi(B)$.
Noting that each limit of linear combinations of $\{X_{t1}, X_{t2}, -\infty < t \le n\}$ and conversely and that $\{z_t\}$ and $\{W_t\}$ are uncorrelated, we see that

$$P_n X_{n+h,2} = \sum_{j=h}^{\infty} \alpha_j z_{n+h-j} + \sum_{j=h}^{\infty} \beta_j W_{n+h-j}$$

The MSE for $t = n + h$ is given by

$$E(X_{n+h,2} - \tilde{P}_n X_{n+h,2})^2 = \sigma_z^2 \sum_{j=0}^{h-1} \alpha_j^2 + \alpha_W^2 \sum_{j=0}^{h-1} \beta_j^2$$

To compute predictors $\tilde{P}_n X_{n+h,2}$, we can rewrite X_{t2} relation with X_{t1} and W_t as

$$A(B)X_{t2} = B^b U(B)X_{t1} + V(B)W_t$$

where A, $\cup$ and V are polynomials of the following form

$$A(B) = 1 - A_1B - \cdots - A_aB_a$$

$$\cup(B) = U_0 + U_1B + \cdots + U_uB^u$$

$$V(B) = 1 + V_1B + \cdots + U_vB_v$$

Applying the operator $\tilde{P}_n$ to transfer function at $t = n + h$, we obtain

$$\tilde{P}_n X_{n+h,2} = \sum_{j=1}^{a} A_j \tilde{P}_n X_{n+h-j,2} + \sum_{j=0}^{u} U_j \tilde{P}_n X_{n+h-b-j,1}$$

$$+ \sum_{j=h}^{v} V_j W_{n+h-j}$$

where the last sum is zero if $h > v$.

As $\{X_{t1}\}$ is uncorrelated with $\{W_t\}$ we predict $X_{n+h-b-j,1}$ as univariate series $\{X_{t1}\}$.

Since n is large, we can replace $\tilde{P}_n X_{j1}$ for each j by the finite past predictor obtained by univariate prediction. The values W_j, $j \leq n$ are replaced by their estimated values $\hat{W}_j$ for least squares estimation as in step (vii) of modelling procedure. We can now solve transfer function equations recursively in order to obtain predictors $\tilde{P}_n X_{n+1,2}$, $\tilde{P}_n X_{n+2,2}$, $\tilde{P}_n X_{n+3,2}$, ... etc.

Example 6
Sales with Leading Indicator

Let
$$X_{t1} = (1 - B)Y_{t1} - .0228, t = 1, \ldots 149$$

$$X_{t2} = (1 - B)Y_{t2} - .420, t = 1, \ldots 149$$

where $\{Y_{t1}\}$ and $\{Y_{t2}\}$, $t = 0, \ldots, 149$ leading indicators and sales data, respectively.

They have zero-mean with models $X_{t1} = (1 - .474B)z_t$, $\{z_t\} \sim WN(0, .0779)$.

We can whiten the $\{X_{t1}\}$ series by applying filter $\hat{\pi}(B) = (1 - .474B)^{-1}$.

Apply $\hat{\pi}(B)$ to both $\{X_{t1}\}$ and $\{X_{t2}\}$, we obtain

$$\hat{z}_t = (1 - .474B)^{-1}X_{t1}, \quad \hat{\sigma}_z^2 = .0779$$

$$\hat{Y}_t = (1 - .474B)^{-1}X_{t2}, \quad \hat{\sigma}_Y^2 = 4.0217$$

The sample cross-correlation function $\hat{\rho}_{Yz}(h)$ of $\{\hat{z}_t\}$ and $\{\hat{Y}_t\}$ shows that $\hat{\rho}_{Yz}(h)$ are non-significant for $h < 3$. Computing $\hat{\tau}_j = \hat{\rho}_{Yz}(j)\hat{\sigma}_Y / \hat{\sigma}_z$ is geometrically decreasing for $h \geq 3$; we have $T(B)$ to be of the form

$$T(B) = w_0(1 - v_1 B)^{-1}B^3$$

Preliminary estimates of w_0 and v_1 are given by

$$\hat{w}_0 = \hat{\tau}_3 = 4.86$$

and $\quad \hat{v}_t = \hat{\tau}_4 / \hat{\tau}_3 = 0.698$

The estimated noise sequence is obtained as

$$\hat{N}_t = X_{t2} - 4.86B^3(1 - .698B)^{-1}X_{t1}$$

The noise sequence $\{\hat{N}_t\}$ has MA(1) model:

$$N_t = (1 - .364B)W_t, \{W_t\} \sim WN(0, 0.0590)$$

So, the preliminary noise cum transfer function models give

$$X_{t2} = 4.86B^3(1 - .698B)^{-1}X_{t1} + (1 - .364B)W_t, \{W_t\} \sim WN(0,0.0590)$$

We minimize the sum of squares in step (vii) with respect to the parameters $(w_0, v_1, \theta_1^{(N)})$ to obtain the least squares model

$$X_{t2} = 4.717B^3(1 - .724B)^{-1}X_{t1} + (1 - .582B)W_t, \{W_t\} \sim WN(0,0.0486)$$

where $X_{t1} = (1 - .474B)z_t, \{z_t\} \sim WN(0,.0779)$.

The least squares model has a reduced noise variance of $\{W_t\}(= 0.0486)$ compared to the preliminary model which had noise variance $(= .0590)$. The sample acf of the series $\{N_t\}$, where

$$N_t = X_{t2} - 4.717B^3 (1 - .724B)^{-1}X_{t1}$$

where only $\rho_1 (= - .582)$ is statistically significant which suggests an MA(1) model for the noise process. The residuals of $\hat{W}_t$ pass the diagnostic tests for white noise, as is required.

The sample cross-correlations between the residuals $\hat{W}_t$ and $\hat{z}_t$, $t = 4,..., 129$ are all non-significant at 5% level.

We can use these results for prediction purposes.

We have $X_{147.1} = -.093$, $X_{149.2} = .08$, $\hat{W}_{149} = -.0706$, $\hat{W}_{149} = .1449$.
Then we find

$$\hat{W}_{149}X_{151.2} = .724\tilde{P}_{149}X_{150.2} + 4.717X_{148.1} + .421W_{149} = .923$$

In terms of the original sales data $\{Y_{t2}\}$, we have $Y_{149.2} = 262.7$ and

$$Y_{t2} = Y_{t-1.2} + X_{t2} + .420$$

Hence, predictors for actual sales are

$$P^*_{149}Y_{150.2} = 262.70 - .228 + .420 = 262.89$$

$$P^*_{149}Y_{151.2} = 262.89 + .923 + .420 = 264.23$$

where P^*_{149} is based on $\{1, Y_{01}, Y_{02}, X_{s1}, X_{s2}, -\infty < s \leq 149\}$ and it is assumed that Y_{01} and Y_{02} are uncorrelated with $\{X_{s1}\}$ and with $\{X_{s2}\}$.
The MSE are given by

$$E(Y_{149+h,2} - P_{149} Y_{149+h,2})^2 = \sigma_z^2 \sum_{j=0}^{h-1} \alpha_j^{*2} + \sigma_w^2 \sum_{j=0}^{h-1} \beta_j^{*2}$$

where

$$\sum_{j=0}^{\infty} \alpha_j^{*2} z^j = 4.717 z^3 (1 - .474z)(1 - .724z)^{-1}(1 - z)^{-1}$$

and

$$\sum_{j=0}^{\infty} \beta_j^{*2} z^j z^j = (1 - .582z)(1 - z)^{-1}$$

For $h = 1$ and 2; we obtain

$$E(Y_{150,2} - P^*_{149}Y_{150.2})^2 = .0486$$

and

$$E(Y_{151,2} - P^*_{149}Y_{151.2})^2 = .0570$$

in good agreement with finite-past MSEs.

It is interesting to examine the improvement obtained by using the transfer function model rather than fitting a univariate model to the sales data alone. Using a univariate model, we obtain $X_{t2} - .249\, X_{t-1,2} - 199\, X_{t-2,2} = U_t$

where $U_t \sim WN(0, 1.794)$ and $X_{t2} = Y_{t2} - Y_{t-1,2} - .420$

The corresponding predictions are 263.14 and 263.58 with MSEs 1.794 and 4.593, respectively. These MSEs are far worse than those obtained for the transfer function model. Hence, transfer function models may be better in a sense of multivariate forecasting rather than univariate forecasting.

4.3.4 Intervention Model Analysis

An intervention model analysis is useful in case of either level changes in the time series which may arise out of policy changes in economic time series or due to faulting (unconformities) in geological/stratigraphic/ drilling data. We assume that the time T, at which the change (intervention) occurs is known, is true in economic/engineering policy change data, but not always true in geological data where most often T is to be estimated from the data. Intervention analysis has same form as transfer functions and was introduced by Box and Tiao (1975) as

$$Y_t = \sum_{j=0}^{\infty} \tau_j X_{t-j} + N_t$$

where the input series $\{X_t\}$ was a deterministic function of t. The mean of Y_t will be $\sum_{j=0}^{\infty} \tau_j X_{t-j}$ and the coefficients $\{\tau_j\}$ are chosen such that the changing level of observations of $\{Y_t\}$ is well represented by the sequence $\sum_{j=0}^{\infty} \tau_j X_{t-j}$.

If $\{Y_t\}$ has $EY_t = 0$ for $T \le t$ and $EY_t \to a \ne 0$ as $t \to \infty$ a suitable input series is

$$X_t = H_t(T) = \sum_{k=T}^{\infty} I_t(k) = \begin{cases} 1 & \text{if } t \ge T \\ 0 & \text{if } t < T \end{cases}$$

Intervention of data gives the time (T_1, T_2, etc.) or locations (L_1, L_2, etc.), at which intervention occurred and suitable input series can then be constructed, such that linear filter coefficients $\{\tau_j\}$ can be estimated by a regression in which errors $\{N_t\}$ constitute an ARMA process. The intervention model then can be used for forecasting and control, etc. As in the case of transfer functions, once $\{X_t\}$ has been chosen (usually permanent level change at T), estimation of the linear filter $\{\tau_j\}$ is simplified by approximating the operator $\sum_{j=0}^{\infty} \tau_j B^j$ with a rational operator of the form

$$T(B) = B^b W(B)/V(B)$$

where b is the delay parameter and $W(B)$ and $V(B)$ are polynomials of the form

$$W(B) = w_0 + w_1 B + \cdots + w_q B^q$$
$$V(B) = 1 - v_1 B - \cdots - v_p B^p$$

By suitable choice of parameters b, p, q and coefficients w_i and v_j, the intervention term $T(B)X_t$ can be made to take a great variety of functional forms.

For example, if $T(B) = wB^2/(1 - vB)$ and $X_t = I_t(T)$, the intervention term is given by

$$\frac{wB^2}{(1 - vB)}\ I_t(T)\ =\ \sum_{j=0}^{\infty} v^j w I_{t-j-2}(T)$$

$$=\ \sum_{j=0}^{\infty} v^j w I_t(T + 2 + j)$$

a series of pulses of sizes $v^j w$ at times $T + 2 + j$, $T = 1, 2...$

If $|v| < 1$, the effect of intervention is to add a series of pulses with size w at time $T + 2$ decreasing to zero at a geometric rate depending on v as $t \rightarrow \infty$.

Similarly, with $X_t = H_t(T)$, we have

$$\frac{wB^2}{(1 - vB)}\ H_t(T)\ =\ \sum_{j=0}^{\infty} v^j w H_{t-j-2}(T)$$

$$=\ \sum_{j=0}^{\infty} (1 + v + \cdots + v^j) w I_t(I + 2 + j)$$

a series of pulses of sizes $(1 + v + \cdots + v^j)w$ at times $T + 2 + j$, $j = 0, 1, 2, \ldots.$

If $|v| < 1$, the effect of the intervention is to bring about a shift in the level series X_t, the size of the shift converging to $w/(1 - v)$ as $t \rightarrow \infty$.

We choose an X_t, which is appropriate and also b, p, q by inspection of the data and then estimate parameters of the rational operators $T(B)$. The model for $\{N_t\}$ is carried out by following steps (vi) to (ix) as in transfer function modelling.

Example 7

Oil Embargo in November 1973

The oil embargo resulted in high oil prices and reduction in the levels of car travelling. Hence, the number of deaths in car accidents was much reduced from zero level (prior to November 1973) to -500 level after November 1973.

These observations were made quarterly from 1970 onwards till 1985 and we, therefore, have

$$X_t = \begin{cases} 0, & \text{if } 0 \le t \le 15 \\ 1, & \text{if } 16 \le t \le 60 \end{cases}$$

The signal data ($t \le 15$) has no structure and is a white noise series with mean set ($=0$) and for $16 \le t \le 60$ we obtain mean $= -486$.

The entire data is accordingly used to obtain the residual series $\{N_t\}$ which has a MA(4) model , i.e.

$$N_t = z_t + .576\, z_{t-1} + .520\, z_{t-2} + .420\, z_{t-3} - .523\, z_{t-4}$$

where $\{z_t\} \sim WN(0,73673)$.

Then, least squares estimations of parameters yielded

$$Y_t = -483\, X_t + N_t$$

where $\quad \{X_t\} = \sum_{k=T}^{\infty} I_t(k) = \begin{cases} 1 & \text{if } t \ge T = 16 \\ 0 & \text{if } t < T = 16 \end{cases}$

and $\quad N_t = z_t + .609\, z_{t-1} + .477\, z_{t-2} + .358\, z_{t-3} - .369\, z_{t-4}$

where $\{z_t\} \sim WN(0,103560)$.

The residuals for N_t pass through whiteness tests (acf and pacfs are all non-significant). So, the intervention model is accepted for use and forecasting future deaths due to car accidents.

4.4   FORECASTING TECHNIQUES

4.4.1   Introduction

We have used the stationary as well as non-stationary parametric models for forecasting time series data such that the forecasts have minimum MSE. Provided the observed series is generated by the proposed and accepted parametric model generated from the time series data, the forecasts provide minimum mean-square errors. However, at times we do not know the true generating process for the time series data and often there are competing parametric models which are equally useful and acceptable and we are not sure which of the model forecasts would, in fact, be correct (and best). Therefore, we should have forecasting techniques that are independent of construction of parametric models from the data and select one that is optimal, according to specific criteria. We cover three general forecasting techniques which are not explicitly dependent on model building and which can be applied to a wide range of real data sets. These three techniques are:

(i) ARAR algorithm, which applies memory-shortening transformations to the data and then fit ARMA model to transformed series;

(ii) Holt-Winters algorithm, which generalizes the experimental smoothing recursives to generate forecasts of time series having a locally linear trend; and

(iii) Holt-Winters seasonal algorithm, which extends (ii) above to handle both trend and seasonality (of known period).

4.4.2 ARAR Algorithm

Given a data set $\{Y_t, t = 1, 2, ..., n\}$, the first step is to decide whether the underlying process is 'long-memory' and, if so, to apply a memory-shortening transformation before we can attempt to fit an AR model. There are two types of memory-shortening differencing operators permitted for ARAR modelling, namely, AR(1) or AR(2) models:

L or AR(1): $\quad \tilde{Y}_t = Y_t - \hat{\phi}(\tilde{\tau})Y_t - \hat{\tau}$

M or AR(2): $\quad \tilde{Y}_t = Y_t - \hat{\phi}_1 Y_{t-1} - \hat{\phi}_2 Y_{t-2}$

We can decide whether a model is (i) L: long-memory and use AR(1), or (ii) M: moderately long-memory and use AR(2) or (iii) S: short memory.

The following steps are followed:

1. For each $\tau = 1, 2, ..., 15$ we find values of $\hat{\phi}(\tau)$ of ϕ that minimizes

$$\mathrm{ERR}(\phi, \tau) = \frac{\sum_{t=\tau+1}^{n}\left[Y_t - \phi Y_{t-\tau}\right]}{\sum_{t=\tau+1}^{n} Y_t^2}$$

We then define $\mathrm{Err}(\tau) = \mathrm{ERR}(\hat{\phi}(\tau), \tau)$ and choose a lag $\hat{\tau}$ to be the value of τ that minimizes $\mathrm{Err}(\tau)$.

2. If $\mathrm{Err}(\hat{\tau}) \leq 8/n$, go to long-memory modelling or AR(1)

3. If $\hat{\phi}(\hat{\tau}) \geq .93$ and $\hat{\tau} > 2$, go to long-memory model or AR(1) or L.

4. If $\hat{\phi}(\hat{\tau}) \geq .93$ and $\tau = 1$ or 2, determine the values of $\hat{\phi}_1$ and $\hat{\phi}_2$ that minimizes

$$\sum_{t=3}^{n} [Y_t - \phi_1 Y_{t-1} - \phi_2 Y_{t-2}]^2$$

then go to moderately long-memory modelling AR(2), i.e. M.

5. If $\hat{\phi}(\hat{\tau}) < .93$, go to short-memory modelling (S).

Next, we fit a subset autoregression as follows:

Let $\{S_t, t = k + 1, ..., n\}$ denote memory-shortened series derived for $\{Y_t\}$ by the algorithm above and let $\overline{S}$ denote the sample means of $S_{k+1}, ..., S_n$.

The next step in modelling procedure is to fit an AR process to the mean-corrected series

$$X_t = S_t - \overline{S}, t = k + 1, ..., n$$

The fitted model is of the form

$$X_t = \phi_1 X_{t-1} + \phi_{l_1} X_{t-l_1} + \phi_{l_2} X_{t-l_2} + \phi_{l_3} X_{t-l_3} + z_t$$

where $\{z_t\} \sim WN(0, \sigma^2)$ and for lags l_1, l_2, l_3 the coefficients ϕ_j and white noise variance σ^2 are obtained by solving the Yule-Walker equations,

$$\begin{bmatrix} 1 & \hat{\rho}(l_1 - 1) & \hat{\rho}(l_2 - 1) & \hat{\rho}(l_3 - 1) \\ \hat{\rho}(l_1 - 1) & 1 & \hat{\rho}(l_2 - l_1) & \hat{\rho}(l_3 - l_1) \\ \hat{\rho}(l_2 - 1) & \hat{\rho}(l_1 - l_2) & 1 & \hat{\rho}(l_3 - l_2) \\ \hat{\rho}(l_2 - 1) & \hat{\rho}(l_3 - l_1) & \hat{\rho}(l_3 - l_2) & 1 \end{bmatrix} \begin{bmatrix} \phi_1 \\ \phi_{l_1} \\ \phi_{l_2} \\ \phi_{l_3} \end{bmatrix} = \begin{bmatrix} \hat{\rho}(1) \\ \hat{\rho}(l_1) \\ \hat{\rho}(l_2) \\ \hat{\rho}(l_3) \end{bmatrix}$$

and $\sigma^2 = \hat{\gamma}(0)[1 - \phi_1 \hat{\rho}(1) - \phi_{l_2} \hat{\rho}(l_1) - \phi_{l_2} \hat{\rho}(l_2) - \phi_{l_3} \hat{\rho}(l_3)]$

where $\hat{\gamma}(j)$ and $\rho(j)$, $j = 0, 1, 2, ...$ are the sample autocovariances and autocorrelations of the series $\{X_t\}$. We compute coefficients ϕ_j for each set of lags such that:

$$1 < l_1 < l_2 < l_3 \leq m$$

where m can be chosen to be either 13 or 26.

It then selects the models for which the Yule-Walker estimate of σ^2 is minimum and we get points at the lags, coefficients and white noise variance for the fitted model. A slower procedure is choosing lags ≤ 13 and computing coefficients by the Yule-Walker equations such that Gaussian likelihood of the observations is maximized (this may be avoided).

Once the memory-shortening filter found in the first step has coefficients $\psi_0(=1)$, ψ_1, ..., $\psi_k(k \geq 0)$, then the memory-shortening series can be expressed as

$$S_t = \psi(B) Y_t = Y_t + \psi_1 Y_{t-1} + \cdots + \psi k Y_{t-k}$$

where $\psi(B) = 1 + \psi_1 B + \cdots + \psi_k B^k$.

Similarly, if the coefficients of the subset autoregression found in the second step are ϕ_{l_1}, ϕ_{l_2}, ϕ_{l_3}, then the subset AR model for the mean-corrected series $\{X_t = S_t - \bar{S}\}$ is

$$\phi(B)X_t = z_t$$

where $\{z_t\} \sim WN(0, \sigma^2)$ and $\phi(B) = 1 - \phi_1 B - \phi_{l_1} B^{l_1} - \phi_{l_2} B^{l_2} - \phi_{l_3} B^{l_3}$.

Therefore, $\xi(B)Y_t = \phi(1)\bar{S} + z_t$

where $\xi(B) = \psi(B)\phi(B) = 1 + \xi_1(B) + \cdots + \xi_{k+l_3} B^{k+l_3}$.

Assuming the ARMA model for $\xi(B)Y_t = \phi(1)\bar{S} + z_t$ is accepted and $\{z_t\}$ uncorrelated with $\{Y_j, j < t\}$ for each t, we can determine the minimum MSE linear predictors $P_n Y_{n+h}$ of Y_{n+h} in terms of $\{1, Y_1, ..., Y_n\}$ for $n > k + l_3$, from the recursives

$$P_n Y_{n+h} = -\sum_{j=1}^{k+l_3} \xi_j P_n Y_{n+h-j} + \phi(1)\bar{S}, h \geq 1$$

with initial conditions $P_n Y_{n+h} = Y_{n+h}$, for $h \leq 0$.
The MSE of $P_n Y_{n+h}$ is found to be

$$E[(Y_{n+h} - P_n Y_{n+h})^2] = \sum_{j=1}^{h-1} \tau_j^2 \sigma^2$$

where $\sum_{j=0}^{\infty} \tau_j z^j$ is the Taylor expansion of $1/\xi(z)$ is a neighborhood of $z = 0$. Equivalently, the sequence $\{\tau_j\}$ can be found for the recursive

$$\tau_0 = 1, \sum_{j=0}^{n} \tau_j \xi_{n-j} = 0, n = 1, 2, ...$$

4.4.3 Holt-Winters Algorithm

We have observations $Y_1, Y_2, ..., Y_n$ for trend plus noise model and the exponential smoothing recursions allowed us to compute the mean-trend $\hat{m}_t$ at times $t = 1, 2, ... n$.

If the series is stationary, then m_t is constant and the exponential smoothing forecast of Y_{n+h} based on $Y_1, Y_2,..., Y_n$ is

$$P_n Y_{n+h} = \hat{m}_n, h = 1, 2, ...$$

If the data possess a non-constant trend, then a natural generalization of the forecast function is

$$P_n Y_{n+h} = \hat{a}_n + \hat{b}_n h, h = 1, 2, ...$$

where $\hat{a}_n$ and $\hat{b}_n$ are estimates of 'level' a_n and 'slope' b_n of the trend function at time n.

Holt (1957) suggested a recursive scheme to compute $\hat{a}_n$ and $\hat{b}_n$.

Denoting by $\hat{Y}_{n+1}$ the one-step forecast $P_n Y_{n+h}$, we have

$$\hat{Y}_{n+1} = \hat{a}_n + \hat{b}_n.$$

Now, as in exponential smoothing, we suppose that the estimated level at time $n + 1$ is a linear combination of the observed value at time $n + 1$.

Thus, $\hat{a}_{n+1} = \alpha Y_{n+1} + (1 - \alpha)(\hat{a}_n + \hat{b}_n)$

We can then estimate the slope at time $n + 1$ as a linear combination of $\hat{a}_{n+1} - \hat{a}_n$ and the estimated slope $\hat{b}_n$ at time n.

Thus, $\hat{b}_{n+1} = \beta(\hat{a}_{n+1} - \hat{a}_n) + (1 - \beta)\hat{b}_n.$

We can solve $\hat{a}_{n+1}, \hat{b}_{n+1}$ by using initial conditions.

A natural choice is to set $\hat{a}_2 = Y_2$ and $\hat{b}_2 = Y_2 - Y_1.$

Then, we can solve for $\hat{a}_i$ and $\hat{b}_i$ recursively, for $i = 3,..., n$ and predictors $P_n Y_{n+h}$ can then be easily found as $\hat{a}_n + \hat{b}_n h; h = 1, 2,...$

These forecasts however, depend on 'smoothing parameters' α and β, which can be either set arbitrarily (values between 0 and 1) or chosen in a more systematic manner to order to minimize the sums of sequences of one-step errors, $\sum_{i=3}^{n} (Y_t - P_{i-1}Y_t)^2.$

The Holt-Winters forecasting procedure is corrected to a steady-state solution of the Kalman filtering equations for a locally-linear structural model with the observation equation

$$Y_t = M_t + W_t$$

and the state-equation

$$\begin{bmatrix} M_t + 1 \\ B_t + 1 \end{bmatrix} = \begin{bmatrix} 1 & 1 \\ 0 & 1 \end{bmatrix} \begin{bmatrix} M_t \\ B_t \end{bmatrix} + \begin{bmatrix} V_t \\ U_t \end{bmatrix}$$

We define $\hat{a}_n$ and $\hat{b}_n$ to be filtered estimates of M_n and B_n, respectively, i.e.

$$\hat{a}_n = M_{n \mid n} = P_n M_n$$

$$\hat{b}_n = B_{n \mid n} = P_n B_n$$

Then, $$\begin{bmatrix} \hat{a}_{n+1} \\ \hat{b}_{n+1} \end{bmatrix} = \begin{bmatrix} \hat{a}_n + \hat{b}_n \\ \hat{b}_n \end{bmatrix} + \Delta_n^{-1} \Omega_n G'(Y_n - \hat{a}_n - \hat{b}_n)$$

where $G = [1 \ 0]$.

Assuming that $\Omega_n = \Omega_1 = [\Omega_{ij}]_{i,j=1}^2$ is the steady-state solution for this model, $\Delta_n = \Omega_{11} + \sigma_w^2$ for all n, so that we obtain

$$\hat{a}_{n+1} = \hat{a}_n + \hat{b}_n + \frac{\Omega_{12}}{\Omega_{11} + \sigma_w^2} (Y_n - \hat{a}_n - \hat{b}_n)$$

$$\hat{b}_{n+1} = \hat{b}_n + \frac{\Omega_{12}}{\Omega_{11} + \sigma_w^2} (Y_n - \hat{a}_n - \hat{b}_n)$$

Solving the above two equations simultaneously, we get

$$\hat{a}_{n+1} = \alpha Y_{n+1} + (1 - \alpha)(\hat{a}_n + \hat{b}_n)$$

$$\hat{b}_{n+1} = \beta(\hat{a}_{n+1} - \hat{a}_n)(1 - \beta)\hat{b}_n$$

with $\alpha = \Omega_{11}/(\Omega_{11} + \sigma_w^2)$ and $\beta = \Omega_{21}/\Omega_{11}$

These equations coincide with the Holt-Winters recursives for $\hat{a}_{n+1}$ and $\hat{b}_{n+1}$, as desired earlier. Therefore, under the local linear model, Holt-Winters and Kalman filtering steady-state solutions are equivalent.

4.4.4 Holt-Winters Seasonal Algorithm

If the series $Y_1, Y_2, ..., Y_n$ contains both trend as well as seasonality of a known period d, then the forecast function is

$$P_n Y_{n+h} = \hat{a}_n + \hat{b}_n h + \hat{c}_{n+h}, \ h = 1, 2,...$$

where $\hat{a}_n, \hat{b}_n, \hat{c}_n$ are estimates of the trend level a_n; trend slope b_n; seasonal component c_n, at time n.

The values of $\hat{c}_{n+h}, h = 1, 2,...,$ are found for the recursive

$$\hat{c}_{n+h} = \hat{c}_{n+h-a}, h \geq 1$$

while the values of $a_i, b_i, c_i, i = 1, ... n$ are found from the recursives

$$\hat{a}_{n+1} = \alpha(Y_{n+1} - \hat{c}_{n+1-d}) + (1 - \alpha)(\hat{a}_n + \hat{b}_n)$$

$$\hat{b}_{n+1} = \beta(\hat{a}_{n+1} - \hat{a}_n)(1 - \beta)\hat{b}_n$$

$$\hat{c}_{n+1} = \gamma(Y_{n+1} - \hat{a}_{n+1}) + (1 - \gamma)\hat{c}_{n+1-d}$$

A natural set of initial conditions for solving above recursives are

$$\hat{a}_{d+1} = Y_{d+1}$$

$$\hat{b}_{d+1} = (Y_{d+1} - Y_1)/d$$

and $\qquad \hat{c}_i = Y_i - (Y_1 + \hat{b}_{d=1}(i-1)), i = 1, 2, ..., d.$

We can solve the recursives successively for $\hat{a}_i, \hat{b}_i, \hat{c}_i, i = d + 1, ... n$ and predictors $P_n Y_{n+h}$ computed easily.

The forecasts, however, depend upon parameters $\alpha, \beta,$ and γ, which can be either arbitrary in range (0 to 1) or chosen in a more systematic way to minimize sum of squares of one-step errors $\sum_{i=d+2}^{n} (Y_i - P_{i-1}Y_i)^2$ applied to already observed data.

Example 8

Car Accident Deaths

Death due to car accidents may vary seasonally as well as level and slope may change due to policy changes in petroleum import and custom duty changes. We model the data after the oil embargo in 1973 by the Middle East companies and assume seasonality $d = 12$ (monthly) and find the best estimates for $\alpha, \beta,$ and γ subject to minimum MSE for one-step ahead forecasts. The error sums of squares are $\sum_{i=14}^{72} (Y_i - P_{i-1}Y_i)/59.$

Table 4.1 Predicted and observed values for $t = 73$ to 78 for the seasonal Holt-Winters algorithm (car accidents in relation to 1973 oil embargo)

$t = years$	1973	1974	1975	1976	1977	1978
Observed Y_t	7798	7406	8363	8460	9217	9316
Predicted HWS	8039	7077	7750	7941	8824	9326

The predicted values seem to be consistently lower than the corresponding observed values. The root mean square error

$$\left(\sum\nolimits_{h=1}^{6} (Y_{72+h} - P_{72}\, Y_{72+h})^2 / 6 \right)^{1/2}$$ for the seasonal Holt-Winters forecasts is

401, which is more than that obtained by using the ARAR algorithm (yielded root mean square estimate of 253) but is better than non-seasonal Holt-Winters algorithm, which gave a value of 1143 and ARIMA models $\nabla_{12} X_t = (0,1,1)X(0,1,1)_{12}$, which gave the root mean square estimate values of 583 and 501 for the same observational data. Therefore, ARAR model forecasts are the best for these data set based on root mean square estimates criterion forecast function. However, it is not necessary that ARAR will prove to be the best forecasts for all types of data sets.

4.4.5 Model-free Forecasting Algorithm

Real data generated by natural processes such as many earth sciences observational data, are rarely if ever, generated by simple mathematical model such as ARIMA or SARIMA processes. Forecasting methods that are predicted on the assumptions of such a model are, therefore, not necessarily the best, even in the MSE sense. Linear additive measurement error model is not always the best error model. Even if minimum MSE for forecasts is followed, we may have to minimize the forecast error variances for each lag separately rather than averaging over several lags. Therefore, heuristic forecasting algorithms discussed in this section may be very useful from a practical forecasting viewpoint. But which of these three algorithms is the best and it is the best for all types of natural historical data?

In addition to additive seasonal models discussed for above three algorithms, we can use multiplicative seasonal model such as $Y_t = m_t\, s_t\, z_t$, where m_t, s_t, and z_t are trend, S_t seasonal and noise factors, respectively. So, it would be useful to transform Y_t into $\ln Y_t$ data where multiplicate model would become linear additive model (such as the linear Holt-Winters) for forecasting processes.

The original values can be obtained by taking anti-logarithms of the forecasts. But ARAR is memory-shortening and hence it gives reasonable forecasts when applied to original (non-linear) data series (sometimes better than forecasts by linear model on $\ln Y_t$ data). This indicates that forecasting methods are not uniquely best for different data.

4.5 SOME CASE STUDIES FROM ECONOMIC GEOLOGY

Geological Example of Time Series Analysis of Borehole Assays

Agucha lead-zinc mine in Rajasthan belongs to Hindustan Zinc Ltd. and is one of the largest opencast mine of lead and zinc in the world. The reserves are 60.36 MT with 13.65% Zn and 1.5% Pb up to a depth of 375 m from the surface. Unfortunately, the assay values are analysed at irregular intervals along any drill which make time series analysis and modelling very difficult. Therefore, borehole assay data have been interpolated at regular intervals to facilitate time series computations and modelling. Polynomials are very useful for interpolation at regular intervals, but real data are not represented properly since it may have jumps/steps and other discontinuities. Therefore, piecewise cubic spline approximation are used for modelling. Cubic spline are piecewise continuous, possess continuous first derivative in interval $[a, b]$, where $a = \tau_1 < \cdots < \tau_n = b$ and no data is lost since $g(x_i) =$ data at knot i. In the interval $[x_i, x_{i+1}]$, the piecewise cubic interpolant f is required to coincide with polynomial P_i of degree three, where ith polynomial P_i is made to satisfy conditions $P_i(x_i) = g(x_i)$ and $P_i(x_{i+1}) = g(x_{i+1})$ apart from the existence of first derivative.

In order to compute coefficients of ith polynomial piece, the equation is

$$P_i(x) = c_{1,i} + c_{2,i} (x - \tau_i) + c_{3,i} (x - \tau_i)^2 + c_{4,i} (x - \tau_i)^3$$

with $c_{1,i} = P_i(\tau_i) = g(\tau_i)$ (assay value) and

$$c_{2,i} = s_i$$
$$c_{3,i} = [\{r_h\ \tau_{i+1}\}g - s_i]/\Delta\tau_i - c_{4,i}\ \Delta\tau_i$$
$$c_{4,i} = s_i + s_{i+1} - 2\,[\tau_i\ r_{i+1}]g\,/(\Delta\tau_i)^2$$

where $\Delta\tau_i = \tau_{i+1} - \tau_i$.

The second coefficient $c_{2,i} = s_i$ is assumed to be zero for $i = 1$ and $i = n + 1$, i.e. $c_{2,1} = 0.0$ or the first slope is taken to be zero. s_i's can be computed as:

$$s_{i-1} \Delta \tau_i + s_i \, 2(\Delta \tau_{i-1} + \Delta \tau_i) \, s_{i+1} (\Delta \tau)_{i-1} = b_i \qquad \qquad ...(1)$$

with $b_i = 3 \{\Delta \tau_i [\tau_{i-1}, \tau_i] g + \Delta \tau_{i-1} + [r_i, r_{i-1}] g\}$, $i = 1, 2, ..., n$.

Equation (1) results in a tridiagonal linear system with $(n-1)$ equations and $(n-1)$ unknowns so can be solved. Once s_1 and s_{n+1} is chosen to be zero, we can solve equation (1) by Gauss elimination. The assays of Zn and Pb were interpolated at 1 m intervals in the boreholes for the time series analysis and modelling. The assay values of zinc that are interpolated with 1 m lags are then transformed to obtain Normal distribution using log $(c/(1-c))$ transformation, where c is fractional assay values. The transformed zinc assay values for 92 BHNO RA 34 with Lat S100 Inc 28 E 105 is given below. The results show the minimum variance occurs when these data differenced once for Zn (as well as for Pb). Therefore, an ARIMA $(p, 1, q)$ model would be suitable for modelling Zn or Pb assay distribution (decreasing linearly down) in this borehole. The acf and pacf for Zn assay distribution show no cut-offs for the acf and pacf, but ρ_1 and ϕ_{11} are statistically significant. Thus, we obtain an ARIMA(1,1,1) model for Zn distribution with $\phi = 0.96$ and $\theta = 0.74$ with $\sigma^2 = 0.10$. Similar analysis for Pb distribution suggested ARIMA(1,1,1) model for the borehole with $\phi = -0.5203$, $\theta = 0.2652$ and $\sigma^2 = 0.06$.

Genesis of Agucha Zinc-Lead Deposit

In the Agucha Zinc-Lead deposit in Rajasthan, the ores are metamorphosed along with host rocks into gneisses and schists belonging to upper amphibolite facies. The ore lenses decrease in assay values for top to bottom (at depth) giving linear trends in assay distribution along depth. Also, since the acf and pacf do not show any cut-offs, it was shown that the model could be ARIMA(1,1,1) with the stationary model ARMA(1,1). We can consider this observed superposed ARIMA(1,1,1) model due to the superposition of two geological processes, that original chemical precipitation of zinc and lead sulphides as ore beds and lenses which was white noise, WN, process or random nucleation/precipitation. These ore beds (lenses) were then subjected to tectonic processes and metamorphosed to amphibolite grades, which was an autoregressive processes (AR(1)) because of diffusion/migration of sulphide molecules from the local source ore lenses (beds) which were at the higher temperature regions at bottom of ore beds towards the top of that sulphide ore bed (lense) simultaneously giving linearity of processes along the depth or ARIMA(1,1,0). Migration of sulphide molecules across the ore beds to higher ore beds were restricted because of low permeability/diffusivity of intervening shale (now schists) horizons which were also carbonaceous in places (now graphite schists).

Therefore, superposition of sedimentation WN process or ARIMA (0,0,0), metamorphic ARIMA(1,1,0) processes has given rise to observed superposed ARIMA(1,1,1) process. This superposed statistical time series model is proved by use of Granger's lemma in proving the superposition of stationary and non-stationary processes.

Modelling of Assays in Mochia Mines, Zawar Mines, Rajasthan

The Mochia mine is in the northern part of Zawar mines and forms an important mine in India producing zinc and lead hosted by dolomites. The associated quartzite and conglomerates and shales (now schists) indicate shallow water deposition of these beds widening the host rocks (dolomites). The associated lead and zinc sulphides were deposited by synsedimentary processes along these sediments (and were not formed by hydrothermal processes). The main reason for this conclusion is the positive association of lead and zinc sulphides, which indicate that these were chemically transported along beds of dolomites but deposited at places under suitable reducing conditions in the shallow epicontinental seas. The association of stromalites also confirms the shallowness of the basin and host rocks (dolomites).

The assay values of Pb, Zn, Cu, Cd, Ag in the sulphide ores were analysed along different drill holes at 1 m intervals which is amenable to time series (spatial series) modelling. These assays were Normalized by $\log(c/(1-c))$ transformation before time series computations. These normalized values were checked for stationarity by differencing and minimum variance of series occurred when $d = 0$, indicating that the series along the drill holes are already stationary for modelling purposes. Based on acf, pacf, the models accepted and verified (by residuals being white noise Normal ($N(0, \sigma^2)$) are as follows:

Table 4.2   Models for different elements on basis of normalized (log C/ (1 – C)) amongs and d = 0

Levels/ Sublevels	Elements & Models	Model accepted
6	Cd, Cu, As, Pb, Zn—all AR(1)	AR(1) for all
XIV	Cd AR(1), Cu AR(1) As AR(1)/AR(2), Pb AR(1)	Cd AR(2); AR(1) for others
XV	Cd AR(1)/AR(2), Cu AR(1), As AR(1) , Zn AR(1)/AR(2)	Cd AR(2); Cu, Ag AR(1); Zn AR(2)

All the chemical components of Pb-Zn sulphides ores are modelled by autoregressive processes (AR(1) and AR(2)), which indicates strong source control in transport and deposition of these sulphide ores along the sedimentary bedding planes. This is possible by transport of sulphide solutions by waves and currents to organic rich reducing environments where these could be deposited along with the dolomitic host rocks (more oxidizing conditions). The presence of stromatolites and dolomites indicate very shallow epicontinental seas for the formation of these sulphide ores under locally reducing conditions.

Later on, some tectonic activities have folded these ores along with the host rocks resulting in metamorphism of sediments, i.e. shales → schists, sandstones → quartzites, and mild alteration to yield sericites/ kaolinites.

Time Series Modelling of Pb-Zn Ores of Rajpura-Darbia Mines, Rajasthan

Rajpura-Darbia mines occur to the south of Agucha Zn-Pb mines and belong to the same geological set-up. It is being mined by underground methods as the lodes are steeply dipping veins, stringers in dolomitic and carbonaceous schist host rocks. The dolomites have some stromalite, preserved at places.

The irregularly spaced borehole assays of Pb, Zn and Ag were regularized at equal intervals of 0.5 m by cubic spline fraction and assay fraction were normalized and closure-corrected by prior $\log(c/(1 - c))$ transform before time series computations (Sahu, 1982). The Table 4.3 and 4.4 give a summary of the results obtained by Koley (1992).

Table 4.3   BH No. RDN/200/1. d = 1 suggests nonstationarity of models

d/Variance of	Zinc	Lead	Silver	Remark
0	0.8007	1.7616	1.4533	Minimum variance at $d = 1$,
1	**0.1223**	**0.3404**	**0.9301**	so first differencing of the series
2	0.2221	0.6024	2.2808	gives stationarity

Table 4.4   Accepted models for bore hole No.s RDN

Zinc	ARIMA(2,1 0)	200/1, 200/19, 200/33, 200/44, 200/45, 209/45, 209/35, 250/7
Lead	ARIMA(2,1,0)	200/1, 200/11, 200/19, 200/33, 200/44, 200/45, 250/3, 250/7
Silver	ARIMA(1,1,1)	200/1, 200/19, 200/33, 200/44, 200/45, 209/35, 250/3, 250/7

Therefore, we explain the genesis of this sulphide ore deposit by above three accepted models with parameter values as follows:

Table 4.5 **Parameters for accepted nonstationary models for the elements**

	Rate	ϕ_1	ϕ_2	θ
Zinc	ARIMA(2,1,0)	.1263	.1449	–
Lead	ARIMA(2,1,0)	– .1966	– .1922	–
Silver	ARIMA(1,1,1)	.3757	–	.6378

An AR(2) model for zinc and lead indicates that there is strong source control over two lags (at ½ m lags). This is interpreted as synsedimentary deposition of Pb-Zn sulphides under deeper marine conditions in reducing environment in the carbonaceous shales and impure limestone/dolomites. The sources of sulphides were deep oceanic furrows, fumaroles, black smokers and carbonates and clays from continents. Later, metamorphism up to amphibolite facies has metamorphosed the ores as well as the host rocks (to garnetiferous graphite schists/gneisses, calc-silicate rocks, etc.). This later tectonism and associated metamorphism have also remobilized the sulphides in hydrothermal veins and stringers gave linear trends, which can be made stationary by using the first differenced series.

Since the silver distribution in the marine exhalations was very small, it was deposited by random nucleation as a white noise process GWN(0, σ^2) which, on remobilization (AR(1)), gave rise to the superposed process ARMA(1,1). Tilting of these veins and stringers gave the linear trend, which can be made stationary by differencing the assays once.

In addition, 18 major and trace components of ores for 22 independent random samples were factor analysed to yield four common factors explaining about 77% of total variance. Varimax rotation of these four common factors and their rotated (varimax) loadings can be used to interpret the sulphide genesis as follows:

Rotated factor 1 (16.42%) has strong positive loadings on Na, Fe and negative loadings on Cu, Mg, Ca, Ag, Mn. This is interpreted as sulphides of copper and silver deposited in limestones and dolomites but not in impure limestones. Rotated factor 2 (17.63%) has strong positive loadings on K, Fe, Ti, Ni. This is interpreted as deposition of limestones and carbonaceous clays in shallow and deeper water environments, respectively. Later metamorphism has given garnetiferous graphitic schists and gneisses. Rotated factor 3 (15.02%) has strong positive loadings on Pb, Zn, Ag, Cd and negative loadings on Si. Base metal sulphides were deposited in deeper marine reducing conditions, whereas influx of carbonates and silicates into these regions gave the gangue materials of these ores.

Rotated factor 4 (17.37%) has strong positive loadings on Ca, Ag and negative loadings on Al, P, Mn, Th, Ni, V. This factor may be interpreted as silver sulphide deposition in limestone hosts during metamorphic remobilization/hydrothermal activity whereas aluminous clays were influx from the continental sources.

Therefore, the time series modelling of borehole assays and factor analysis of the ores complement the interpretations in a coherent manner regarding sedimentary deposition of marine exhalative sulphide-rich solutions in deeper waters and shallow water sedimentation of sandstones, dolomites and impure limestones nearer to the continents. Later tectonic activities and associated metamorphism of these rocks and ores gave rise to some minor remobilization of ores and hydrothermal deposition as cavity filling and replacement ores.

Time Series Analysis of Base Metal (Cu, Pb, Zn) Deposits of Kolihan mines, Rajasthan

Copper, zinc and lead sulphide deposits of Khetri copper belt, Rajasthan, occur within the schists and amphibolite quartzites belonging to Ajabgarh Group (Delhi Supergroup) as veins, stringers and are cofolded with the host rocks. This indicates that synsedimentary sulphides were deposited as beds, lenses primarily in the shales and secondarily in argillaceous sediments, which on folding and metamorphism, have been converted to garnetiferous chlorite schists and amphibole quartzites, respectively. During the tectonic and metamorphic activity, some redistribution and mobilization of sulphides in the form of hydrothermal solution has occurred and given rise to veins and with replacement texture deposits and cavity fillings in voids and fractures of host rocks, as well. This hydrothermal activity has also produced wall rock alterations in the form of chloritization, sericitization, silification, etc.

Borehole assay data were regularized to equal intervals by the use of cubic spline interpolation method (de Boor, 1978. Sahu and Bagchi, 1993), which is piecewise continuous. The boreholes in garnetiferous schists host rock are deposits gave models which were ARIMA(1,1,0), indicating superposed AR(1) processes with strong source syndepositional control, whereas models in amphibolite quartzite host rock deposits were ARIMA(2,1,0), indicating superposed processes of initial AR(1) source syndepositional control with later metamorphism/hydrothermal diffusions AR(1) process.

Table 4.6   Stationarity for bore hole assays for copper at lags 1 m

d	Borehole number	
	UKG – 82	UKG – 75
0	.1010	.1106
1	**.0780**	**.0742**
2	.1780	.1991
3	.5158	.6438

Table 4.7   Accepted models for ores in amphibole quartzites

Boreholes	Model	ϕ_1	ϕ_2	σ^2
UKF -33	ARIMA(1,1,0)	.1495	–	0.04796
UKG-82	ARIMA(1,1,0)	– .1235	–	0.09589
UKG-8	ARIMA(1,1,0)	– .1472	–	0.07562
UKG-46	AR(1)	.9678	–	0.37726
UKM-12	AR(1)	.9671	–	0.37720
UKC-38	ARIMA(2,1,0)	– .5648	– .4517	0.07284
UKF-44	ARIMA(2,1,0)	.2738	– .2330	0.02335

Table 4.8   Accepted models for ores graniteferous schists (Lodes, II), Lag = 1m

Boreholes	Model	ϕ	θ	σ^2
UKG-48	ARIMA(1,0,1)	.6125	– .8605	.592100
UKG-2	ARIMA(0,1,1)	–	– .5313	.043646
UKG-75	ARIMA(0,1,1)	–	– .4200	.063600
UKG-49	ARIMA(0,1,1)	–	– .3050	.043230

Factor analysis of 19 normalized major and trace elemental components of ores using 22 independent random samples gave 4 common component factors using the scree test and second order criteria for cumulative covariances of principal components (Sahu, 1973). The rotation of these four common factors by varimax has identified the following common factors for genetic interpretation of sulphide ores based on rotated loadings and eigenvalues.

Rotated factor 1 is bipolar with strong positive loadings on Na_2O, Al_2O_3, CaO, K_2O, TiO_2, Cr and strong negative loadings on MnO, Fe_2O_3, indicating intense weathering and alteration of country rocks due to hydrothermal and/or weathering activities with the production of clay minerals like sericite, montmorillonite, illite containing K_2O, CaO, Na_2O, Al_2O, TiO_2 etc. and having some MnO, and Fe_2O_3.

Rotated factor 2 is also bipolar with strong positive loadings on P_2O_5, Zn, Cu, Ni, Fe_2O_3 and very strong negative loadings on SiO_2. This can be interpreted as remobilization and hydrothermal sulphide mineralization of Cu, Zn in silica-poor veins and stringers (mesothermal conditions).

Rotated factor 3 has positive loadings on MgO, TiO_2, Pb and negative loadings on P_2O_5, Cu. This indicates that galena and copper sulphides were remobilized and deposited at separate parts of the veins and stringers whereas Zn sulphides were not remobilized (epithermal conditions).

Rotated factor 4 indicates an alteration factor of amphiboles and garnets where Ni, Co replace Mg but not Ca.

Geological Problems Associated with Banded Haematite Quartzite Ores

The iron ores (BHQ) in Goa are worked as an opencast mine at Bicholim, Goa. This mine was sampled at 1 m intervals vertically and 3 m intervals along strike (horizontally) for the time (spatial) series modelling. In addition to the marketable component of hematite (Fe_2O_3) or total iron (Fe) in the ores, there were two deleterious components, silica (SiO_2) and phosphorous (as P_2O_5), which greatly reduced the quality of iron ore and its market price. Therefore, multivariate time series modelling of all these components were necessary for grade control and marketing purposes as well as for decision as regards necessity for blending and/or ore beneficiation. In addition, it was hoped that these times series models would help in deciphering the genetic processes of deposition/ formation of iron ores, source of iron and conditions of deposition in the basin etc. The following technological questions were of immediate interest and these could be answered by corresponding time (spatial) series modelling and forecasting:

1. What are the models for distribution of Fe_2O_3, SiO_2, P_2O_5, TiO_2, MnO in Bicholim mine along strike (65 samples, 3 m intervals) and along depth (vertically 40 samples, 1 m intervals) of the ore beds? For these purposes, the assay values were normalized by $\log (c/(1 - c))$ transformation before ARIMA(p, d, q) models could be fitted to each series separately (univariate models).

Once the models are accepted, parameters estimated and validated by residuals checking to be white noise, they could be used for technological forecasting as also technological decisions.

2. Multivariate time (spatial) series models were used with single input and multiple inputs. These models would give better technological forecasts and decisions.

3. Multivariate transfer function cum noise models were established with single output and multiple inputs, so that the forecasts could improve in accuracy and precision for better technological decisions as regards blending, beneficiation or marketing.

4. Although multiple outputs and multiple input models could be worked out, this was not done since most of the questions posed could be answered by the above simpler time (spatial) series models.

5. The main geological questions were: (a) Why there are alterations of hematite and silica bands (laminae) in BHQ deposits? (b) What is the source of iron? (c) What was the environment of deposition of BHQ in the basin and (d) tectonic conditions that prevailed? and (e) Were there any post-depositional changes occurring in the ore deposits?

We shall summarize the geological inferences which can be made from time (spatial) series modelling of BHQ ore deposits. All sequences (horizontal along strike, vertical along depth) were stationary and hence ARMA(p, q) models were adequate with $q = 0$ in almost all vertical directions, indicating a strong source/lateritization control for deposition of iron and silica over a period of time. The maximum total Fe is 68% at the top and irregularly decays to about 40% at 40 m below the surface in the mine, which indicates the occurrence of considerable weathering/leaching of silica layers. Thus, the top BHQ horizons have very high grades (lateritization process). In the lateral direction (along strike), there is evidence of oscillation of ore grade every 18 m from west to east, without any specific trends, so there are microbasins of deposition of that scale. Alternate iron and silica layers of deposition occurred due to chemical supersaturation, resulting in the deposition of one component and its dilution in the milieu giving a system of oscillating limit cycles.

Univaraite Time (Spatial) Series Modelling of Components of Iron Ores in Bicholim mines, Goa

In univariate time series modelling of components of iron ores belonging to BHQ of Bicholim mines, Goa, it was found that $\log(c/(1 - c))$ transformation Normalized the data for lateral (along strike) as well as vertical (along depth) series and no differencing was required for any component in order to make the sequence stationary. Therefore, ARMA(p, q) model would be required for these components along the lateral and vertical directions. Based on the acf and pacf values as well as checking the residuals were white noise Gaussian, i.e. WNN(0, σ^2), the following models were finally accepted as all the residuals were found to be non-significant.

Table 4.9   Lateral series (65 samples, 3 m intervals)

−2.04833	SiO_2	AR(1)	$x_t = \bar{x} + .544\, x_{t-1} + z_t$	$\sigma^2 = 0.0458$
1.07417	Fe_2O_3	AR(1)	$x_t = \bar{x} + .358\, x_{t-1} + z_t$	$\sigma^2 = 0.0272$
−3.29196	TiO_2	AR(1)	$x_t = \bar{x} + .413\, x_{t-1} + z_t$	$\sigma^2 = 0.0631$
−2.57602	MnO	AR(1)	$x_t = \bar{x} + .221\, x_{t-1} + z_t$	$\sigma^2 = 0.0245$
−2.80381	P_2O_5	AR(2)	$x_t = \bar{x} + .311\, x_{t-1} + .361 x_{t-2} + z_t$	$\sigma^2 = 0.0216$

Table 4.10   Vertical samples (40 samples, 1 m intervals)

−2.00048	SiO_2	AR(1)	$x_t = \bar{x} + .803\, x_{t-1} + z_t$	$\sigma^2 = 0.0313$
0.94930	Fe_2O_3	AR(1)	$x_t = \bar{x} + .443\, x_{t-1} + z_t$	$\sigma^2 = 0.0122$
−3.19654	TiO_2	AR(1)	$x_t = \bar{x} + .477\, x_{t-1} + z_t$	$\sigma^2 = 0.0153$
−2.39846	MnO	AR(1)	$x_t = \bar{x} + .486\, x_{t-1} + z_t$	$\sigma^2 = 0.2585$
	P_2O_5	White Noise	$\bar{x} = -2.73858$ $\sigma^2 = 0.0039$	

Subsequent lateritization has resulted in higher grade iron to be concentrated towards the top surface of Bicholim mines. From these models, we have strong source control for Fe_2O_3, SiO_2, etc., in BHQ, which indicate that volcanic emanations in deep-sea floors may have given these components and chemical sedimentation due to supersaturation alternated (loop cycles) so that alternate layers of iron ore (haematite) and silica were deposited as BHQ in the basin.

The following are univariate time series forecasts up to five lead times for the pivotal observations.

Lateral model forecasts: AR(1) or AR(2)

Table 4.11   SiO$_2$ at pivot 63: 1.18%, ϕ_1 = .544, σ^2 = .0458, $t_{0.95,\infty}$ = 2.00, $t_{0.50,\infty}$ = 0.6745

Lead	s.d. of f.error	Observed value	Forecast	+95%	−95%	+50%	−50%
1	0.2142	0.80	1.04	2.73	0.39	1.44	0.75
2	0.3029	0.70	0.97	3.78	0.24	1.54	0.61
3	0.3710	-	0.93	4.92	0.17	1.64	0.52
4	0.4284	-	0.91	6.19	0.13	1.75	0.47
5	0.4790	-	0.90	7.61	0.10	1.87	0.43

Table 4.12   Fe$_2$O$_3$ at pivot 62: 94.44%, ϕ_1 = .358, σ^2 = .0272

Lead	s.d. of f.error	Observed value	Forecast	+95%	−95%	+50%	−50%
1	0.1648	93.30	93.07	96.62	89.34	94.55	91.23
2	0.2331	91.10	92.51	97.31	80.86	94.66	89.59
3	0.2855	93.89	92.31	97.81	76.30	94.92	88.50
4	0.3296	–	92.22	98.16	72.51	95.19	87.67
5	0.3685	–	92.20	98.47	68.40	95.44	86.96

Vertical Model Forecasts: AR(1):

Note: P$_2$O$_5$ cannot be modelled as time series, since it is Gaussian stationary.

Table 4.13   SiO$_2$, AR(1), Pivot 33: 1.40%, ϕ_1 = .803, σ^2 = 0.0313

Lead	s.d. of f.error	Observed value	Forecast	+95%	−95%	+50%	−50%
1	0.1770	1.88	1.30	2.95	0.58	1.72	0.99
2	0.2504	3.43	1.24	3.86	0.39	1.82	0.84
3	0.3066	–	1.18	4.75	0.29	1.90	0.74
4	0.3541	–	1.14	5.66	0.22	1.97	0.66
5	0.3958	–	1.11	6.61	0.18	2.05	0.60

Table 4.14 Fe_2O_3, AR(1), Pivot 33: 91.30%, $\phi_1 = .443$, $\sigma^2 = 0.0122$

Lead	s.d. of f.error	Observed value	Forecast	+95%	−95%	+50%	−50%
1	0.1107	88.90	90.54	94.12	85.12	91.93	88.95
2	0.1565	84.52	90.19	95.01	81.61	92.15	87.80
3	0.1917	-	90.03	95.66	78.73	92.42	86.99
4	0.2213	−	89.96	96.16	76.19	92.68	86.36
5	0.2474	−	89.92	96.58	73.84	92.93	85.83

Multivariate Time Series Models for Components of Iron Ores in Bicholim Mines, Goa

Since the components of iron ores such as SiO_2, Fe_2O_3, TiO_2, MnO, P_2O_5 are themselves autocorrelated, we have to prewhiten the inputs and obtain post-whitened residuals for cross-correlations with the residuals of significant component to be forecasted (in this case, Fe_2O_3). The results for lateral series (along strike of beds) and vertical series (along the depth) are given below.

Table 4.15 Lateral series ($N = 64$)

Fe_2O_3 residuals (y_t) at lags	Post-whitened residuals (x_{it})				Remarks
	SiO_2 (X_{1t})	$TiO_2(X_2t)$	$MnO(X_{3t})$	$P_2O_5(X_{4t})$	$|\gamma|$ at 5% significance is 0.246
$\gamma_{yx}(0)$	− .488*	− .395*	− .438*	− .372*	
$\gamma_{yx}(1)$	− .310*	− .213	.172	0.135	* statistically significant at 5% level
$\gamma_{yx}(2)$	− .047	.200	− .121	− .233	
$\gamma_{yx}(3)$	− .105	− .240	.099	− .059	
$\gamma_{yx}(-1)$	− .220	− .080	− .109	− .200	
$\gamma_{yx}(-2)$	− .023	.038	− .105	− .096	

These results show that we can use SiO_2, TiO_2, MnO, P_2O_5 as additional inputs in order to forecast Fe_2O_3 values and these inputs act instantaneously. However, SiO_2 acts as lag 1 as well and a transfer function model with delay time 1 can also be useful if SiO_2 input series is used for forecasting Fe_2O_3 series. These components can be used for better multivariate forecasts of Fe_2O_3 in lateral series as compared to univariate forecasts of Fe_2O_3 for Fe_2O_3 model.

Multivariate forecast functions of Fe_2O_3 reduces the univariate forecast error variance (σ^2) by 33.2%, which is considerable reduction in errors and consequent increase in precision.

Table 4.16   Vertical series (N = 34)

Fe_2O_3 residuals (y_t) at lags	Post-whitened residuals (x_{it})				Remarks
	$SiO_2(X_{1t})$	$TiO_2(X_{2t})$	$MnO(X_{3t})$	$P_2O_5(X_{4t})$	$|\gamma|$ at 5% significance is 0.246
$\gamma_{yx}(0)$	− .481*	− .757*	− .691*	− .193	
$\gamma_{yx}(1)$	− .006	− .026	− .081	.182	*statistically signifi- cant at 5% level
$\gamma_{yx}(2)$	− .059	.041	− .108	.018	
$\gamma_{yx}(3)$	.222	− .053	− .162	− .163	
$\gamma_{yx}(-1)$	− .220	− .087	− .104	.259	
$\gamma_{yx}(-2)$	.010	− .053	.230	.154	

The cross-correlations at different lags indicate that components act instantaneously and SiO_2, TiO_2 and MnO can be used to obtain better multivariate forecasts of Fe_2O_3 for the Fe_2O_3 model.

This also shows that P_2O_5 information does not help in obtaining better multivariate forecasts for the vertical series, as none of its cross-correlations is significant at the 5% level. Multivariate forecast function for Fe_2O_3 along the vertical section reduces the univariate forecast error variance by 74.8%, which is a very great reduction in error and a converse increase in precision.

Univariate and Multivariate Transfer Function cum-Noise Models for Components of Iron Ores of Bicholim Mines, Goa

Transfer function models establish a relation between the output time series (to be forecasted) and a set of related input random time series through a linear filter. Since the distribution of P_2O_5, in addition to Fe_2O_3 content, plays an important role in value of marketing of iron ores, it was thought prudent to forecast these two components of iron ores by univariate (single input-single output) and multivariate (multiple input-single output) transfer function technology. Since the cross-correlations were maximum at zero lags, there were no pure delay processes and the transfer function models were considerably simple as

$$Y_t = \delta^{-1}(B) \, w(B) \, X_{t-b} + N_t$$

where delay parameter b was found to be zero.

The impulse response function

$$v_k = \gamma_{\alpha\beta}(k)s_\beta/s_\alpha \text{ for } k = 0, 1, 2, 3 \ldots$$

where $\gamma_{\alpha\beta}(k)$ is the cross-correlation between prewhitened input (β) and prewhitened output (α) series and s_β and s_α are standard deviations of α and β series.

For Fe_2O_3 output series in lateral data, we have

$$\alpha_{1t} = X_{1t} - .544\, X_{1t} - 1 \qquad\qquad (X_{1t} = SiO_2)$$

$$\alpha_{2t} = X_{2t} - .413\, X_{2t} - 1 \qquad\qquad (X_{2t} = TiO_2)$$

$$\alpha_{3t} = X_{3t} - .221\, X_{3t} - 1 \qquad\qquad (X_{3t} = MnO)$$

$$\alpha_{4t} = X_{4t} - .311\, X_{4t-1} - .361 X_{4t-2} \qquad (X_{4t} = P_2O_5)$$

The (lateral) univariate transfer function model were as shown in Table 4.17.

Table 4.17   Lateral univariate transfer function models

		Noise variance	Noise model (N.M)
Fe_2O_3 vs SiO_2	$Y_t = -.319 X_{1t} + N_t$	0.0242	Absent
Fe_2O_3 vs TiO_2	$Y_t = -.250 X_{2t} + N_t$	0.0209	Absent
Fe_2O_3 vs MnO	$Y_t = -.176 X_{3t} + N_t$	0.0243	Present
Fe_2O_3 vs P_2O_5	$Y_t = -.380 X_{4t} - .298 X_{4t-1} + N_t$	0.0255	Present

Similarly, P_2O_5 output models for lateral data were:

P_2O_5 vs Fe_2O_3	$Y_t = -.357\, X'_{2t} + N_t$	0.0268	present
P_2O_5 vs MnO	$Y_t = .159\, X_{4t} + N_t$	0.0255	present

The corresponding univariate transfer function equations for vertical series were as shown in Table 4.18.

Table 4.18   Vertical univariate Transfor function models

Lateral series	Fitted transfer function model	σ^2	N.M.
Fe_2O_3 vs SiO_2	$Y_t = -.347 X_{1t} + N_t$	0.0140	Present
Fe_2O_3 vs TiO_2	$Y_t = -.681 X_{2t} + N_t$	0.0065	White
Fe_2O_3 vs MnO	$Y_t = -.150 X_{3t} + N_t$	0.0066	White
Vertical series			
P_2O_5 vs SiO_2	$Y_t = .139 X_{1t} + N_t$	0.0667	Present

Forecasting by univariate transfer function models by single input-single output case

(a) Lateral series

Table 4.19   Forecast variances and standard deviation for Fe_2O_3

	Lead 1	Lead 2	Lead 3	Lead 4	Lead 5
Forecast variance	0.0209	0.0418	0.0626	0.0835	0.1044
Forecast std deviation	0.1445	0.2044	0.2503	0.2890	0.3231

Input TiO_2 Pivot 62: .038% output: Fe_2O_3

Table 4.20   Forecast and confidence intervals for Fe_2O_3

Forecasted at location/lead	Observation	Transfer function forecast	+95	−95	+50	−50
63/(1)	93.30	92.94	95.94	86.20	93.83	90.66
64/(2)	91.10	92.27	96.83	82.33	94.25	89.68
65/(3)	93.89	92.22	97.40	78.91	94.59	88.93
66/(4)	–	92.20	97.81	75.74	94.87	88.29
67/(5)	–	92.19	98.12	72.71	95.12	87.72

(b) Vertical series

Table 4.21   Brid

	Lead 1	Lead 2	Lead 3	Lead 4	Lead 5
Forecast std deviation	0.0806	0.1139	0.1395	0.1611	0.1801

Input series TiO_2 Pivot 34: TiO_2 = 0.081% output series Fe_2O_3

Table 4.22   Forecast and confidence intervals for Fe_2O_3

Forecasted at location/lead	Observation	Transfer function forecast	+95	−95	+50	−50
35/(1)	84.52	89.00	92.17	84.76	90.13	87.70
36/(2)	–	89.39	93.47	83.21	90.96	87.57
37/(3)	–	89.61	94.29	81.23	91.47	87.39
38/(4)	–	89.73	94.87	80.50	91.83	87.16
39/(5)	–	89.80	95.32	79.20	92.11	86.91

Multiple Input-single Output Transfer Function Forecasting Models

The multivariate regression relation is as follows:

$$b = R^{-1}g$$

where b is vector of regression coefficients, R is correlation of inputs, g is vector of cross-correlation coefficients between prewhitened output and inputs.

(a) Fe_2O_3 output lateral section with inputs SiO_2, TiO_2, MnO, P_2O_5
   We obtain $b_1 = -.127^{n.s.}$, $b_2 = -.281^*$, $b_3 = -.370^*$, $b_4 = 0.008^{n.s.}$
   Eliminating b_1, b_1, we obtain $b_2 = -.341^*$, $b_3 = -.408^*$ with
   $R^2 = 0.310$ (31%).
   So, the final equation Y_t is compared as

$$\frac{Y_t}{s_y} = \frac{-.341}{.27404}X_2 + \frac{-.408}{.42562}X_3 + N_t$$

   where Y_t is estimated Fe_2O_3, X_2 = input TiO_2, X_3 = input MnO,
   $s_y = 0.1648$

(b) Fe_2O_3 output vertical section with inputs SiO_2, TiO_2, MnO
   We obtain $b_1 = -.337^*$, $b_2 = -.374^*$, $b_3 = -.527^*$, $s_y = 0.1107$,
   $R^2 = 0.821$ (82.1%).

$$\text{So, } \frac{Y_t}{s_y} = \frac{-.337}{.28664}X_1 + \left(\frac{-.374}{.14175}\right)X_2 + \left(\frac{-.527}{.58943}\right)X_3 + N_t$$

$$= -1.175X_1 - 2.63845 X_2 - .89408 X_3 + N_t$$

Lateral section forecasts of Fe_2O_3
Pivot 62: TiO_2: 0.038%, MnO: 0.159%, output—Fe_2O_3

Table 4.23 Transfer function forecasts and confidence intervals for Fe_2O_3

Forecasted at location/ lead	Observation	F.F. standard deviation	Transfer function forecast	+50	−50
63/(1)	93.30	.1390	92.51	93.87	90.87
64/(2)	91.10	.1966	92.29	94.20	89.82
65/(3)	93.89	.2408	92.22	94.52	89.08
66/(4)	–	.2781	92.20	94.79	88.97
67/(5)	–	.3109	92.10	95.03	87.93

Vertical section forecasts of Fe_2O_3

Pivot 32: SiO_2: 2.65%, TiO_2: 0.046%, MnO: 0.203%, output—Fe_2O_3

Table 4.24   Transfer function forecasts and confidence intervals for Fe_2O_3

Forecasted at location/ lead	F.F. standard deviation	Observation	Transfer function forecast	+50	−50
33/(1)	.0455	91.30	89.71	90.35	89.04
34/(2)	.0643	88.90	89.51	90.41	88.54
35/(3)	.0788	84.52	89.46	90.56	88.25
36/(4)	.0910	–	89.48	90.74	88.07
37/(5)	.1017	–	89.52	90.92	87.95

Modelling Forecasting Average Grade of Blocks of Gold Ores, Kolar

The following is a purely illustrative example for modelling and average grade forecasting of blocks in the gold ores of Kolar area. The average gold content in 49 blocks is given by Sarma (1979) and we assume that these blocks are of same size and arranged along one line for purposes of this example. Since the content of gold is very small, these within the limit are log-normally distributed (Sahu, 1995) and stochastic modelling is done for log (assay a_i) rather than original assays a_i. The variance of various differenced series of log a_i values clearly shows that log a_i series is stationary and no differencing is required for modelling purposes. This is in contrast to the theory propounded by the French school, who assume that only the first differenced series is stationary and useful for modelling. Using the log a_i series, we compute autocorrelations r_k and partial autocorrelation ϕ_{kk} by standard formulae (Box and Jenkins, 1970, 1976) and conclude that only the ϕ_{11} is statistically significant, indicating an autoregressive first-order process (AR(1)). The equation of this Markov or AR(1) process is easily computed as:

$$\tilde{z}_t = 0.3055\tilde{z}_{t-1} + a_t$$

where $\tilde{z}_t$ is the deviation from mean (2.6971) and a_t is a random shock at position t.

The variance of noise $\sigma_a^2 = 0.04619$ ($\sigma_a = 0.2149$), which is rather high (90.6%). Therefore, the forecasting will not be very good and should not be done beyond three steps or so. The variance for n-step ahead forecasts is given by $\pm \tau_{\infty/50} = 0.6745$, which is to be multiplied by the corresponding n-step ahead forecast variance. The results are tabulated in Table 4.25.

A study of the data presented in the Table 4.25 clearly indicates that the assay values may be stationary by themselves and differencing is not always required for stochastic modelling. Assays are correlated positively in the blocks and reflect an autoregressive first-order (Markov) process. Had the data been available with closer spacing, the autocorrelating could be much higher. Since large block size average out ore-shoots and lean zones, the block assays tend to become almost random (white noise) process. Therefore, point assay data could be better suited for stochastic modelling and forecasting purposes than averaged assay data, over a large volume of ore.

This example shows that stochastic models will be useful for grade forecasts, decision on minable blocks, grade control and blending, calculation of reserves and assays, etc. In addition, the type of models and the values of parameters obtained for the accepted model may throw light on the origin of ores, geological processes and the like which will aid their prospecting and evaluation. For a detailed study of this example, readers are referred to B.K. Sahu (1982).

Table 4.25 **Data for gold assay width (dwt inch for 49 blocks and generation of stochastic forecast function)**

1	425	14	453	27	875	40	473
2	305	15	359	28	308	41	321
3	320	16	479	29	479	42	651
4	382	17	475	30	264	43	2075
5	245	18	446	31	338	44	1951
6	409	19	579	32	374	45	503
7	586	20	616	33	1525	46	425
8	420	21	929	34	290	47	1830
9	409	22	598	35	500	48	1224
10	414	23	228	36	451	49	718
11	493	24	407	37	488		
12	765	25	305	38	370		
13	400	26	233	39	628		

$\log a_i$ series	$\nabla \log a_i$ series	$\nabla_2 \log a_i$ series
$\bar{x}\ 2.6971 = \mu$	0.0049	-0.0019
$s^2\ 0.05095^* = \sigma_a^2$	0.07419	0.20445
$s.\ 0.22573 = \sigma_a$	0.27238	0.45216

*minimum variance; hence stationary and no differencing is required, (d = 0).

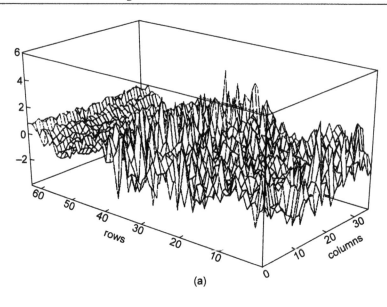

(a)

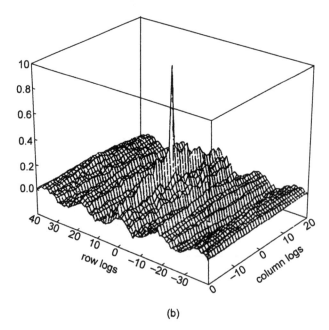

(b)

Fig. 4.1 (a) 2-D time series of soil-temperature data taken along a rectangular field
(b) 2-D autocorrelation function for the soil temperature data

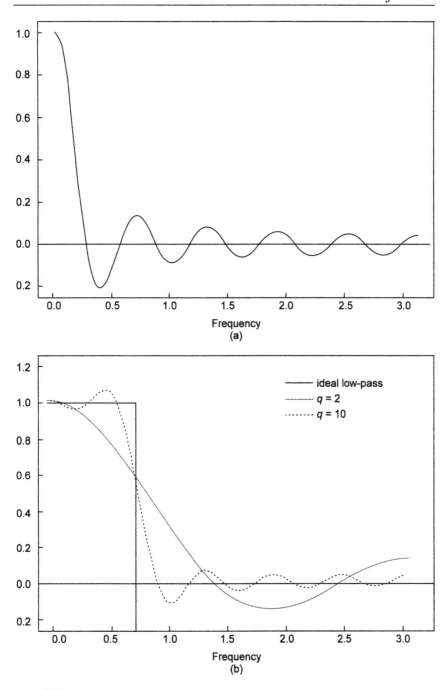

Fig. 4.2 (a) Transfer function of a simple MA model
(b) Transfer function of low-pass filters

Autocorrelations and partial autocorrelations for log a_i series are needed for modelling and are found to be 0.3065 which is significantly only for ϕ_{11}. Forecasting is done by adding the mean value of 2.6971 to the predicted $\tilde{z}_t$ and then taking the antilogarithm.

Table 4.26   Autocorrelation as partial autocorrelation for log a_i.

For log a_i series	Lag1	Lag2	Lag3	Lag4	Lag5	Lag6
r_k	$0.3055^{n.s.}$	$0.0077^{n.s.}$	$0.0730^{n.s.}$	$0.2582^{n.s.}$	$0.1460^{n.s.}$	$0.0837^{n.s.}$
ϕ_{kk}	0.3055^*	$-0.0144^{n.s.}$	$0.0376^{n.s.}$	$-0.0132^{n.s.}$	–	–

The forecast function model is Markov or AR(1) process $\tilde{z}_t = 0.3055$ $\tilde{z}_{t-1} + a_t$.

Variance of forecast errors at different leads and standard deviations are given as below:

Table 4.27   Variance and std. deviation of forecast errors at different leadtimes

	Lead1	Lead2	Lead3	Lead4	
Variance	0.04619	0.09238	0.13857	0.18476	$t_{50,\infty} = 0.6745$
Std deviation	0.2149	0.3039	0.3723	0.4298	

Forecasts with different intervals computed from the above model are shown in Table 4.28.

4.6 EXERCISES

1. Let $\{X_t\}$ be the bivariate time series where components are MA(1)

   $$X_{t1} = z_{t,1} + .8\, z_{t-1,1}, \{z_{t1}\} \sim \text{IID}\ (0, \sigma_1^2)$$
   $$X_{t2} = z_{t,2} - .6\, z_{t-1,2}, \{z_{t2}\} \sim \text{IID}\ (0, \sigma_2^2)$$

   where $\{z_{t1}\}$ and $\{z_{t2}\}$ are independent. Find large sample approximation to the covariance of $n^{1/2}\hat{\rho}_{12}(h)$ and $n^{1/2}\hat{\rho}_{12}(k)$ for $h \neq k$.

2. Show that multivariate best linear predictor of Y in terms of $\{X_1, \ldots, X_n\}$ is orthogonal to one-step prediction errors $X_j - \hat{X}_j$ and $X_k - \hat{X}_k$ for $h \neq k$.

3. Determine the covariance matrix function of ARMA(1,1) process satisfying

   $$X_t - \Phi X_{t-1} = z_t - \Theta z_{t-1}, \{z_t\} \sim \text{WN}(\mathbf{0}, \mathbf{I}_2)$$

Table 4.28 Forecasts and confidence intervals for Gold arrays of different Blocks from Kolar Gold Mine

Pivot 47: value 1830				*Pivot 48: value 1224*				*Pivot 49: value 718*			
Lead	Forecast	+50%	−50%	Lead	Forecast	+50%	−50%	Lead	Forecast	+50%	−50%
48	741	1035	531	49	655	915	469	50	557	777	399
49	562	1015	311	50	541	977	300	51	515	930	285
50	517	921	290	51	511	910	286	52	503	897	282
51	504	981	255								

where I_2 is 2×2 identity matrix and $\Phi = \Theta' = \begin{bmatrix} 1/2 & 1/2 \\ 0 & 1/2 \end{bmatrix}$.

4. If $\{X_t\}$ is as causal AR(p) process such that

$X_t = \Phi_1 X_{t-1} + \cdots + \Phi_p X_{t-p} + z_t, \{z_t\} \sim WN(0, S)$

write the recursives for predictors $P_n X_{n+h}, h \geq 0$ for $n \geq p$.

Give error covariance matrix in terms of ϕ_i and Σ for $h = 1, 2, 3$.

5. Let $\{Y_t\}$ be a multivariate ARIM ($p, 1, 0$) process such that $\nabla Y_t = X_t$ (of problem 4 above). Assuming $E(Y_0 X_1') = 0$, for $t \geq 1$, show that

$$\tilde{P}_n(Y_{n+h}) = Y_n + \sum_{j=1}^{h} P_n X_{n+j}.$$

Derive error factors for $h = 1, 2, 3$.

6. Give the formula for MSE of h-step forecast based on ARAR algorithm as $E\left[(Y_{n+h} - P_n Y_{n+h})^2\right] = \sum_{j=1}^{h-1} \tau_j^2 \sigma^2$.

7. In the Holt-Winters algorithm, show that $\alpha = \Omega_{11}/(\Omega_{11} + \sigma_w^2)$ and $\beta = \Omega_{21}/\Omega_{11}$ for estimating level a_n and slope b_n.

References

1. Box, G.E.P. and Tiao, G.C., 1975, Intervention analysis with applications to economic and environmental problems, *J. Amer. Stat. Assocn.*, 70: 70-79.

2. deBoor, C.A. 1978, *A practical guide to splines*, New York, Springer Verlag.

3. Hannan, E.J., 1970, *Multiple Time Series*, Wiley, New York.

4. Hannan, E.J., and Deistler, M., 1988, *Statistical Theory of Linear Systems*, Wiley, New York.

5. Holt, C.C., 1957, Forecasting Seasonals and Trends by Exponentially Weighted Moving Averages, *ONR Res. Memo.*, 52, Carnegie Inst. Pittsburgh. P.Q., USA.

6. Jones, R.H., 1978, Multivariate Autoregression Estimation using Residuals, In: D.F. Findley (Ed.) *Applied Time Series Analysis*, Academic Press, New York, pp. 139-162.

7. Koley, M. 1992, M.Tech. Dissertation, IIT, Bombay.

8. Lütkepohl, H., 1993, *Introduction to Multiple Time Series Analysis* (2nd Edition), Springer, Berlin.

9. Raiker, P.S., 1982, Ph.D., IIT Bombay.

10. Raiker, P.S., and Sahu, B.K., 1985 Single/multiple input-output transfer function-cum-noise models for constituents of iron ores of Northern Goa, India, *Mathematical Geology*, 17: 755-767.

11. Sahu, B.K., 1982, Stochastic Modelling of Mineral Deposits, *Mineral Deposita*, 17: 99-105.

12. Sahu, B.K., and Raiker, P.S., 1985 Univariate and multivariate stochastic modelling of iron ores of Northern Goa, India, *Mathematical Geology*, 17: 317-325.

13. Sahu, B.K., and Bogchi, J. 1993, Statistical Modelling of lead zinc deposit at Rampura Agucha, Rajasthan. *Jour. Geol. Soc. India*, 41: p. 231-242.

14. Sahu, B.K., 1995, Statistical inference from geochemical and petrographic data, *Proc. Rec. Res. Geol. of Western India*, Baroda.

15. Sarma, D.D., 1979, *A Statistical Appraisal of Ore Evaluation*, Andhra Univ. Press, Waltiar, p. 167.

ADVANCED TIME SERIES MODELS

5.1 INTRODUCTION

In this chapter, we discuss a few more advanced and powerful modelling techniques for time series data analysis, forecasting and control. The most attractive features (see Jones, 1984; Tong, 1983) are:

(i) Observations are not required to be made at equal time intervals

(ii) Missing data poses no problems

These models have not yet been applied to geological data analysis but have great potential for use in future. We cover state-space models, non-linear models, ARCH and GARCH models and long-memory models, some of which are very useful for earth sciences data analysis in time and/or frequency domains.

Suppose we have p multiple variables y_{ti}, which are related to inputs x_{ti} plus residual z_{ti}, which are correlated with i but uncorrelated with time.

The $y_{ti} = \beta_{i1}x_{t1} + \beta_{i2}x_{t2} + \cdots + \beta_{iq}x_{tq} + z_{ti}$

for each $i = 1, 2, ..., p$ and $j = 1, 2, ..., q$

We assume Cov $(z_{is}, z_{jt}) = \sigma_{ij}$ for $s = t$ and zero otherwise.

Then, in matrix notation,

$$y_t = Bx_t + z_t$$

The MLE of regression matrix $\hat{B} = Y'X(X'X)^{-1}$ and error covariance matrix Σ is estimated as

$$\hat{\Sigma}_z = \frac{1}{(n-q)} \sum_{t=1}^{n} (y_t - \hat{\beta}x_t)(y_t - \hat{\beta}x_t)'$$

The standard errors for estimates is given as

$$Se((\hat{\beta}_{ij})) = \sqrt{\hat{\sigma}_{jj} \, c_{ii}}$$

where $\hat{\sigma}_{jj}$ is the jth diagonal element of $\hat{\Sigma}_z$ and c_{ii} is the ith diagonal

element of $\left(\sum_{t=1}^{n} x_t \, x_t' \right)^{-1}$

Also the information criterion is given by

$$AIC = \ln \left| \hat{\Sigma}_z \right| + \frac{z}{n} \left(pq + \frac{p(p+1)}{2} \right)$$

$$SIC = \ln \left| \hat{\Sigma}_z \right| + \left(pq + \frac{p(p+1)}{2} \right) \frac{\ln n}{n}$$

$$AICC = \ln \left| \hat{\Sigma}_z \right| + \left(\frac{p(q+m)}{m - p - q - 1} \right) \text{(for a multivariate case)}.$$

The multivariate AR model is given by VAR(1)

$$x_t = \alpha + \Phi \, x_{t-1} + z_t$$

where Φ is a $p \times p$ transition matrix giving dependence of x_t on x_{t-1}. The vector white noise z_t is expressed as

$$E \, z_t \, z_t' = \Sigma_z$$

and vector $\alpha = (\alpha_1, \alpha_2, ..., \alpha_p)'$ are constants of regression.
If $E(x_t) = \mu$, then $\alpha = (I - \phi)\mu$

The MLE of $\hat{\Sigma}_z = (n-1)^{-1} \sum_{t=2}^{n} (x_t - \hat{\alpha} - \hat{\phi}x_{t-1})(x_t - \hat{\alpha} - \hat{\phi}x_{t-1})'$

VAR(1) can be extended to VAR(p^*) as

$$y_t = \alpha + \sum_{k=1}^{p^*} \phi_k \, x_{t-k} + z_t$$

$$\text{Res SSQCP} = \sum_{t=p+1}^{n} (y_t - \hat{\beta}x_t)(p^*y_t - \beta x_t)'$$

Schwartz criterion SIC $= \ln \left| \hat{\Sigma}_z \right| + \dfrac{p^2 p^* \ln n}{n}$ gives better results than AIC or AICC criterion for order selection. In ARMAX, we also include some exogenous vector variables (u_t) to influence the outputs. The model for x_t is then

$$x_t = \Gamma u_t + \sum_{j=1}^{p^*} \Phi_j x_{t-j} + \sum_{k=1}^{p^*} \Theta_k w_{t-k} + w_t$$

where Γ is a $p \times r$ parameter matrix and u_t is $r \times 1$ fixed vector of inputs. This does not create any problem as Γu_t can be estimated as intercept (α) in multivariate regression model.

In recent years, state-space representations with Kalman filtering have gained importance in control of linear systems and are an extremely rich class of models well beyond linear ARIMA/SARIMA or classical decomposition models. For example, the trend and seasonal components are allowed to evolve randomly, data with missing values can be modelled and Bayesian priors can also be imposed.

In non-linear time series modelling, we have four main categories as:

(i) bilinear

(ii) random coefficient AR

(iii) threshold AR

(iv) exponential AR

All these categories can be combined into one model (v) termed State Dependent Model (SDM). These non-linear models are much useful in several geological processes such as earthquakes, chemical deposition of ore minerals, evolution of species, etc. ARMAX models have an external influencing factor which has to be accounted for.

ARCH (p) and GARCH (p, q) models are developed to capture the changing volatility of a few time series data such as stock-returns. Long-memory models are useful for hydrological data analyses and are ARIMA(p, d, q) processes with $0 < |d| < \frac{1}{2}$ in contrast to ARMA(p, q) processes, which have short memory (i.e. acf $\rightarrow 0$ after a few lags).

5.2 STATE-SPACE MODELS

5.2.1 State-Space Representation

For state-space representation, we use the following notation throughout this section:

$$\{W_t\} \sim \text{WN}(0, \{R_t\}) \qquad \qquad ...(5.2.1)$$

to indicate that the random vectors W_t have mean 0 and

$$E(W_s W_t') = \begin{cases} R_t & \text{if } s = t \\ 0 & \text{otherwise} \end{cases} \qquad ...(5.2.2)$$

A state-space model for a multiple time series $\{Y_t, t = 1, 2, ...\}$ consists of 'observation equation', expressing w-dimensional observations plus corresponding noise. Thus,

$$Y_t = G_t X_t + W_t, \quad t = 1, 2, ... \qquad ...(5.2.3)$$

where $\{W_t\} \sim WN(0, \{R_t\})$ and $\{G_t\}$ is a sequence of $w \times v$ matrices.

A second equation for state-space representation is the 'state equation', which determines the state X_{t+1} at time $t + 1$ in terms of state X_t and a noise term. The state equation is

$$X_{t+1} = F_t X_{t-1} + V_t, \quad t = 1, 2, ... \qquad ...(5.2.4)$$

where $\{F_t\}$ is a sequence of $v \times v$ matrices, $\{V_t\} \sim WN(0, \{Q_t\})$ and $\{V_t\}$ is uncorrelated with $\{W_t\}$. We assume that the initial state X_1 is uncorrelated with $\{V_t\}$ and $\{W_t\}$, which completes the state-space specification.

In addition, in control theory, we have an additional term $H_t u_t$ in state equation to apply control 'u_t' at time t for influencing X_{t+1}. We can also relax the uncorrelated ness of V_t and W_t for a more general state-space representation.

The matrices F_t, G_t, Q_t, and R_t may be independent of t in most applications (i.e. these are time constant matrices) and hence, the subscript t may be dropped for the model. X_t and Y_t have the following functional form for $t = 2, 3, ...$

$$X_t = F_{t-1} X_{t-1} + V_{t-1}$$

$$= F_{t-1}(F_{t-2} X_{t-2} + V_{t-2}) + V_{t-1}$$

$$= \cdots = f_t (X_1, V_1, ..., V_{t-1}) \qquad ...(5.2.5)$$

and $\qquad Y_t = g_t (X_1, V_1, ..., V_{t-1}, W_t) \qquad ...(5.2.6)$

where $E(V_t X_s') = 0$; $E(V_t Y_s') = 0$, $1 \le s \le t$ and $E(W_t X_s') = 0$; $1 \le s \le t$

We can have state-space representation for a large number of time series and other models, where $\{X_t\}$ and $\{Y_t\}$ are not necessarily stationary. The beauty of state-space representation lies in its simple structure for state equation, which permits very simple analysis for the process $\{X_t\}$. The behaviour of $\{Y_t\}$ is then easy to determine from that of $\{X_t\}$, using the observation equation (5.2.3). If the sequence $\{X_1, V_1, V_2, ...\}$ is independent, then $\{X_t\}$ has the Markov property, i.e. distribution X_{t+1} given $X_t, ..., X_1$ is same as distribution of X_{t+1} given X_t.

If we include a sufficiently large number of components in the specification of the state X_t (for example, X_t includes components of X_{t-1} for each t), we will have Markov property. Akoi, 1987, may be referred for more details.

Example 1

Give state-space representations for AR(1), ARMA(1,1).

(i) AR(1):

Let $\{Y_t\}$ be causal AR(1) process $Y_t = \phi Y_{t-1} + z_t$, $\{z_t\} \sim$ WN$(0, \sigma^2)$.

So, $\quad X_{t+1} = \phi X_t + V_t$, $t = 1, 2,...$ $\qquad$ ...(5.2.7)

where $\quad X_1 = Y_1 = \sum_{j=0}^{\infty} \phi^j z_{i-j}$ and $\gamma_t = z_{t+1}$.

The process $\{Y_t\}$ then satisfies the observation equation $Y_t = X_t$, where $G_t = 0$ and $W_t = 0$.

(ii) ARMA (1,1):

Let $\{Y_t\}$ be causal-invertible ARMA(1, 1) process given by

$$Y_t = \phi Y_{t-1} + z_t + \theta z_{t-1}, \{z_t\} \sim \text{WN}(0, \sigma^2).$$

The state-space representation for $\{Y_t\}$ is given by

$$Y_t = \theta(B)X_t = [0,1]\begin{bmatrix} X_{t-1} \\ X_t \end{bmatrix}$$

where $\{X_t\}$ is the causal AR(1) process satisfying $\phi(B)X_t = z_t$ or equivalent equation

$$\begin{bmatrix} X_t \\ X_{t+1} \end{bmatrix} = \begin{bmatrix} 0 & 1 \\ 0 & \phi \end{bmatrix}\begin{bmatrix} X_{t-1} \\ X_t \end{bmatrix} + \begin{bmatrix} 0 \\ z_{t+1} \end{bmatrix}$$

Since $X_t = \sum_{j=0}^{\infty} \phi^j z_{t-j}$, then the above equation forms the state-space representation of $\{Y_t\}$ with

$$X_t = \begin{bmatrix} X_{t-1} \\ X_t \end{bmatrix} \text{ and } X_t = \begin{bmatrix} \sum_{j=0}^{\infty} \phi^j z_j \\ \sum \phi^j z_{1-j} \end{bmatrix}$$

Extensions to general ARMA and ARIMA processes will be given later.

State-space models are much more versatile than many other models. A more natural formulation of time series models is time index that runs from $-\infty$ to $+\infty$ and we should modify the state-space representation accordingly. We then have observation and state equations,

$$Y_t = GX_t + W_t, t = 0, \pm 1, \ldots \qquad \ldots(5.2.7)$$

$$X_{t+1} = FX_t + V_t, t = 0, \pm 1, \ldots \qquad \ldots(5.2.8)$$

where F and G are $v \times v$ and $w \times v$ matrices, respectively, $\{V_t\} \sim WN(0, Q)$, $\{W_t\} \sim WN(0, R)$ and $E(V_s, W_t') = 0$ for all s and t.

Equation (5.2.8) is stable if matrix F has all its eigenvalues in the interior of the unit circle or equivalently, if $\det(I - F_z) \neq 0$ for all z complex such that $|z| \leq 1$. The matrix F is then said to be stable. If (5.2.8) have the unique stationary solution as

$$X_t = \sum_{j=0}^{\infty} F^j V_{t-j-1}$$

The corresponding observations $Y_t = W_t + \sum_{j=0}^{\infty} GF^j V_{t-j-1}$ is also stationary.

5.2.2 Structural Model

A structural time series model, such as classical decomposition, is specified in terms of components: trends, seasonality and noise, which are of interest themselves. However, in the classical decomposition model, we assume the trend and seasonality to be deterministic, which limits the applicability. Under state-space representation, we assume that trend and seasonality are randomly varying and estimation and forecasting are possible through Kalman recursions (see 5.2.4).

Example 2   *Random walk plus noise (non-seasonal model)*

We have $Y_t = M_t + W_t$, where $\{W_t\} \sim WN(0, \sigma_w^2)$.

Here, we allow randomness in the 'local level' via random walk noise V_t, so that $M_t = m_t$ and further $M_{t+1} = M_t + V_t$ and $\{V_t\} \sim WN(0, \sigma_v^2)$ with initial value $M_1 = m_1$.

The difference data D_t is given by

$$D_t = \nabla Y_t = Y_t - Y_{t-1} = V_{t-1} + W_t - W_{t-1}, t \geq 2$$

is a stationary time series with mean 0 and acf

$$\rho_D(h) \; = \; \begin{cases} -\sigma_w^2/(2\sigma_w^2 + \sigma_v^2) & \text{if } |h|=1 \\ \quad\quad 0 & \text{if } |h|>1 \end{cases}$$

Since $\{D_t\}$ is 1-correlated, we have an MA(1) model for $\{D_t\}$ and hence $\{Y_t\}$ is an ARIMA(0,1,1) process.

So, $\quad\quad\quad\quad D_t \; = z_t + \theta z_{t-1}, \; \{z_t\} \sim \text{WN}(0, \sigma^2)$

where θ and σ^2 are found by solving the following equations:

$$\theta/(1 + \theta)^2 \; = - \sigma_w^2/(2\sigma_w^2 + \sigma_v^2) \text{ and } \theta\sigma^2 = - \sigma_w^2$$

Suppose that $\theta/(1 + \theta)^2 = -.4$ and $\theta\sigma^2 = -8$, we find that $\theta = -\frac{1}{2}$ and $\sigma^2 = 16$ (the other solution $\theta = -2$ and $\sigma^2 = 4$ is not admissible).

In this model, the local level m_t changes over a period of time t and measures the characteristic output of an identical process for which the unobserved process level $\{M_t\}$ is intended to be within specified limits (to meet design and marketing requirements). To decide whether or not the process require any corrective action (or control), we test that process level $\{M_t\}$ is constant from the state equation. We see that $\{M_t\}$ is constant (equal to m_1) when $V_t = 0$ or when $\sigma_v^2 = 0$. This means that MA(1) model for $\{D_t\}$ is non-invertible with $\theta = -1$. We can also test for unit-root hypothesis for parameter $\theta = -1$.

The local level model can be easily extended to incorporate the locally-linear trend with slope β at time t. Then,

$$M_t \; = M_{t-1} + B_{t-1} + V_{t-1},$$

where $B_{t-1} = \beta_{t-1}$.

We now introduce randomness into the slope by replacing it with the random walk

$$B_t \; = B_{t-1} + U_{t-1}$$

where $\{U_t\} \sim \text{WN}(0, \sigma_u^2)$ to obtain the 'locally linear trend model'. In state-space form, the 'locally linear trend model' is

$$X_t \; = (M_t, B_t)'$$

The $\quad\quad X_{t+1} \; = \begin{bmatrix} 1 & 1 \\ 0 & 1 \end{bmatrix} X_t + V_t, t = 1, 2, \ldots \quad\quad\quad\quad \text{... (5.2.9)}$

where $V_t = (V_t, U_t)'$.

The process $\{Y_t\}$ is then determined by the observation equation

$$Y_t = [1 \ 0] \ X_t + W_t \qquad \qquad ...(5.2.10)$$

If $\{X_t, U_1, V_1, W_1, U_2, V_2, W_2,...\}$ is an uncorrelated sequence, the equations (5.2.9) and (5.2.10) constitute a slope-space representation of process $\{Y_t\}$, which is a model for randomly varying trend and added noise. For this model, we have $v = 2$, $w = 1$.

$$F = \begin{bmatrix} 1 & 1 \\ 0 & 1 \end{bmatrix}, G = [1 \ 0], Q = \begin{bmatrix} \sigma_v^2 & 0 \\ 0 & \sigma_u^2 \end{bmatrix} \text{ and } R = \sigma_w^2$$

Example 3

(i) *AR(1) with noise*

The univariate causal AR(1) process with noise in state-space is given by

$$y_t = x_t + v_t$$

where the signal x_t satisfies $x_t = \phi x_t + w_t$

The noise processes are independent and Gaussian with $\sigma_w = \sigma_v = 1$.

Let the state process be AR(1) with $\phi = 0.8$, but we observe y_t not x_t. The acvf of x_t is

$$\gamma_x(h) = [\sigma_w^2 / (1 - \phi^2)] \phi^h, h = 0, 1, 2,....$$

The observations are stationary since y_t is the sum of two independent stationary processes $\{x_t\}$ and $\{v_t\}$. So, acvf of y is

$$\gamma_y(0) = \text{Var} \ (y_t) = \text{Var} \ (x_t + v_t) = \frac{\sigma_w^2}{1 - \phi^2} + \sigma_v^2$$

and when $h \geq 1$, we have

$$\gamma_y(h) = \text{Cov} \ (y_t, y_{t-h}) = \gamma_x(h).$$

So, acf for y process for $h \geq 1$ is

$$\rho_y(h) = \gamma_y(h) / \gamma_y(0) = \left(1 + \frac{\sigma_v^2}{\sigma_w^2} (1 - \phi^2) \right)^{-1} \phi^h$$

Thus, observation processes y_t are not AR(1) unless $\sigma_v^2 = 0$ and acf of y_t is the same as for ARMA(1,1).

Therefore, $y_t = \phi y_{t-1} + \theta u_{t-1} + u_t$

where u_t is Gaussian white noise with variance σ_u^2 and θ and σ_u^2 are to be suitably chosen. Although stationary ARMA and stationary state-space models are equivalent, it is easier to work with state-space models for missing data, complex multivariate systems, mixed effects and certain types of non-stationarities.

(ii) *ARMA(1, 1) with noise.*

The univariate ARMA(1,1) process

$$y_t = \phi y_{t-1} + \theta v_{t-1} + v_t$$

can be written in state equation form using ARMAX property as:

$$x_{t+1} = \phi x_t + w_t \text{ (state equation)}$$

where $w_t = (\theta + \phi)v_t$ and $y_t = x_t + v_t$ (observation equation)

In this case, $\mathrm{Cov}(w_t, v_t) = (\theta + \phi)^2$, $\mathrm{Var}\, v_t = (\theta + \phi)^2 R$ and $\mathrm{Cov}(w_t, v_s) = 0$ when $s \neq t$.

The likelihood can be obtained by residuals v_t series (innovations algorithm) as

$$-2 \ln L\, Y(\theta) = \sum_{t=1}^{n} \log |\Sigma_t(\theta)| + \sum_{t=1}^{n} \varepsilon_t(\theta)' \Sigma_t(\theta)^{-1} \varepsilon_t(\theta)$$

where θ includes the parameter vectors (ϕ, θ).

This is highly non-linear, so maximum likelihood can be obtained by Newton-Raphson iterations or by EM algorithm.

Example 4 *Seasonal series with noise*

In classical decomposition of $\{X_t\}$ into trend, noise, seasonality and noise components, we had the seasonal component with period d which was a sequence $\{s_t\}$ such that $s_{t+d} = s_t$ and $\sum_{t=1}^{d} s_t = 0$.

Such a space can be generated for *any* values of $s_1, s_0, \ldots, s_{-d+3}$ by means of recursives

$$s_{t+1} = -s_t - \cdots - s_{t-d+2}, t = 1, 2, \ldots.$$

We now allow for random deviations for strict periodicity by adding a term V_t to the RHS above where $\{V_t\}$ is a white noise with zero mean. This gives recursive relations for observed series $\{Y_t\}$ as

$$Y_{t+1} = -Y_t - \cdots - Y_{t-d+2} + V_t, t = 1, 2, \ldots$$

To make state-space representation for $\{Y_t\}$, we introduce a $(d-1)$-dimensional state vector $X_t = (Y_t, Y_{t-1}, \ldots, Y_{t-d+2})'$

The series $\{Y_t\}$ is then given by the observation equation

$$Y_t = [1\ 0\ 0\ \ldots\ 0\,]\, X_t, t = 1, 2, \ldots.$$

where $\{X_t\}$ satisfies the state equation

$$X_{t+1} = F X_t + V_t, t = 1, 2,...$$

$$V_t = (V_t, 0, 0, ...,0)'$$

and

$$F = \begin{bmatrix} -1 & -1 & \cdots & -1 & -1 \\ 1 & 0 & \cdots & 0 & 0 \\ 0 & 1 & \cdots & 0 & 0 \\ \vdots & \vdots & \ddots & \vdots & \vdots \\ 0 & 0 & \cdots & 1 & 0 \end{bmatrix}$$

Example 5 *Randomly-varying trend with seasonal and noise components*
We add the two series of previous examples. We have a state vector

$$X_t = \begin{bmatrix} X_t^1 \\ X_t^2 \end{bmatrix}$$

where X_t^1 and X_t^2 are the state vectors for two previous examples, respectively.

The state equation for the second series $\{Y_t\}$ is given by

$$X_{t+1} = \begin{bmatrix} F_1 & 0 \\ 0 & F_2 \end{bmatrix} X_t + \begin{bmatrix} V_t^1 \\ V_t^2 \end{bmatrix} \qquad ...(5.2.11)$$

where F_1, F_2 are previous two coefficient matrices and $\{V_t^1\}$, $\{V_t^2\}$ are corresponding noise vectors, respectively.
The observation equation is

$$Y_t = [1\ 0\ 1\ 0\ ...\ 0]\,X_t + W_t \qquad ...(5.2.12)$$

where $\{W_t\}$ is the noise sequence $\sim \mathrm{WN}(0, \sigma_w^2)$.

If the sequence of random vectors $\{X_1, V_1^1, V_1^2, W_1, V_2^1, V_2^2, W_2\}$ is uncorrelated, the equations (5.2.11) and (5.2.12) constitute a state-space representation for $\{Y_t\}$.

Example 6 *Global warming*
The observation series from 1880 to 1987 consists of continental observation (y_{t1}) and marine data (y_{t2}), both of which have same undergoing climatic signal (x_t) but with different noise series (v_{t1}, v_{t2}). The signal model may be a random walk $x_t = x_{t-1} + w_t$. Thus, its first difference series is stable. Therefore, observation equation is

$$\begin{pmatrix} y_{t1} \\ y_{t2} \end{pmatrix} = \begin{pmatrix} 1 \\ 1 \end{pmatrix} x_t + \begin{pmatrix} v_{t1} \\ v_{t2} \end{pmatrix}$$

and covariance matrices are given by Q of DLM $= q_{11}$ and $R = \begin{pmatrix} r_{11} & r_{12} \\ r_{21} & r_{22} \end{pmatrix}$.

So, we can estimate R, Q, which are unknown and keep the transition scheme matrix $\Phi = \begin{pmatrix} 1 \\ 1 \end{pmatrix}$ at fixed level.

5.2.3 State-Space Representation of ARIMA Models

We start for causal AR(p) process and then build up the representations for ARMA and ARIMA processes.

Example 7 *Causal AR(p)*

Consider AR(p) process defined as

$$Y_{t+1} = \phi_1 Y_t + \phi_2 Y_{t-1} + \cdots + \phi_p Y_{t-p} + z_{t-1}, \quad t = 0, \pm 1, \ldots$$

where $\{z_t\} \sim \text{WN}(0, \sigma^2)$ and $\phi(z) = 1 - \phi_1 z - \cdots - \phi_p z^p$ is non-zero for $|z| \le 1$. To express $\{Y_t\}$ in the state-space form, we introduce the state vectors

$$X_t = \begin{bmatrix} Y_{t-p+1} \\ Y_{t-p+2} \\ Y_t \end{bmatrix}, t = 0, \pm 1, \ldots. \qquad \ldots(5.2.13)$$

Then the observation equation is given by

$$Y_t = [0 \ 0 \ 0 \ \ldots \ 1]X_t, t = 0, \pm 1, \ldots \qquad \ldots(5.2.14)$$

and the state equation is given by

$$X_{t+1} = \begin{bmatrix} 0 & 1 & 0 & \cdots & 0 \\ 0 & 0 & 1 & \cdots & 0 \\ \vdots & \vdots & \vdots & \vdots & \vdots \\ 0 & 0 & 0 & \cdots & 1 \\ \phi_p & \phi_{p-1} & \phi_{p-2} & \cdots & \phi_1 \end{bmatrix} X_t + \begin{bmatrix} 0 \\ 0 \\ \vdots \\ 0 \\ 1 \end{bmatrix} z_{t+1}, t = 0, \pm 1, \ldots \qquad \ldots(5.2.15)$$

These equations have the required forms for infinite time sequences with $W_t = 0$ and $V_t = (0, 0, \ldots, z_{t+1})'$, $t = 0, \pm 1, \ldots$

The causality condition $\phi(z) \neq 0$ for $|z| \leq 1$ for AR(p) is equivalent to the condition that the state equation (5.2.15) is stable, since the eigenvalues of coefficients matrix in (5.2.15) are reciprocals of the zeros of $\phi(z)$.

If time $t = 1, 2, \ldots$ and if X_1 is a random vector such that $\{X_1, z_1, z_2, \ldots\}$ is an uncorrelated sequence, then $\{Y_t\}$ has a state-space representation (5.2.1), (5.2.2).

The resulting process $\{Y_t\}$ is well-defined, regardless of whether or not the state equation is stable, but it will not be stationary, in general. Stationarity is brought in if the state equation is stable and if X_1 is defined by (5.2.13) with $Y_t = \sum_{j=0}^{\infty} \psi_j z_{t-j}$, $t = 1, 0, \ldots, 2-p$, and $\psi(z) = 1/\phi(z)$, $|z| \leq 1$.

Example 8 *Causal ARMA(p, q) processes*

State-space representations are not unique for this model and we give one possible representation.

Let ARMA(p, q) is $\phi(B)Y_t = \theta(B)z_t$, $t = 0, \pm 1, \ldots$

where $\{z_t\} \sim$ WN($0, \sigma^2$) and $\phi(z) \neq 0$ for $|z| \leq 1$.

Let $r = \max(p, q+1)$, $\phi_j = 0$ for $j > p$, $\theta_j = 0$ for $j > q$ and $\theta_0 = 1$.

If $\{U_t\}$ is the causal AR(p) process satisfying $\phi(B)U_t = z_t$, then $Y_t = \theta(B)U_t$ since

$$\phi(B)Y_t = \phi(B)\theta(B)U_t = \theta(B)\phi(B)U_t = \theta(B)z_t$$

So, $Y_t = [\theta_{r-1} \ \theta_{r-2} \ \theta_1] X_t$ (observation equation)

where $X_t = \begin{bmatrix} U_{t-r+1} \\ U_{t-r+1} \\ \vdots \\ U_t \end{bmatrix}$

Then, we obtain

$$X_{t+1} = \begin{bmatrix} 0 & 1 & 0 & \cdots & 0 \\ 0 & 0 & 1 & \cdots & 0 \\ \cdots & \cdots & \cdots & \cdots & \cdots \\ 0 & 0 & 0 & \cdots & 1 \\ \phi_r & \phi_{r-1} & \phi_{r-2} & \cdots & \phi_1 \end{bmatrix} X_t + \begin{bmatrix} 0 \\ 0 \\ \vdots \\ 0 \\ 1 \end{bmatrix} z_{t+1}, \ t = 0, \pm 1, \ldots \text{(state equation)}$$

As for causal model AR(p), we obtain the observation and state noise vectors are

$$W_t = 0, \text{ and } V_t = (0, 0, \ldots, z_{t+1})', t = 0, \pm 1, \ldots$$

Example 9 *ARIMA(p, d, q) process*

If $\{Y_t\}$ is an ARIMA(p, d, q) process with $\{\nabla^d Y_t\}$ satisfying ARMA(p, q), then $\{\nabla^d Y_t\}$ has the representation

$$\{\nabla^d Y_t\} = GX_t, \ t = 0, \pm 1, \ldots$$

where $\{X_t\}$ is the unique stationary solution of the state equation

$$X_{t+1} = FX_t + V_t$$

where F, G are unique coefficients of X_t as in ARMA(p, q) process and $V_t = (0, 0, \ldots, z_{t+1})'$.

Let A and B be the $d \times 1$ and $d \times d$ matrices defined by $A = B = 1$ if $d = 1$ and

$$
A = \begin{bmatrix} 0 \\ 0 \\ \vdots \\ 0 \\ 1 \end{bmatrix}, B = \begin{bmatrix} 0 & 1 & 0 & \cdots & 0 \\ 0 & 0 & 1 & \cdots & 0 \\ \cdots & \cdots & \cdots & \cdots & \cdots \\ 0 & 0 & 0 & \cdots & 1 \\ (-1)^d \binom{d}{d} & (-1)^d \binom{d}{d-1} & (-1)^d \binom{d}{d-2} & \cdots & d \end{bmatrix} \text{ if } d > 1
$$

Then, since $Y_t = \nabla^d - \sum_{j=1}^{d} \binom{d}{j}(-1)^j Y_{t-j}$ the vector $Y_{t-1} = (Y_{t-d}, \ldots, Y_{t-1})'$

satisfies the equation

$$Y_t = A \nabla^d Y_t + BY_{t-1} = AGXt + BY_{t-1}$$

Defining a new state vector T_t by stacking X_t and Y_{t-1}, we obtain the state equation

$$T_{t+1} = \begin{bmatrix} X_{t+1} \\ Y_t \end{bmatrix} = \begin{bmatrix} F & 0 \\ AG & B \end{bmatrix} T_t + \begin{bmatrix} V_t \\ 0 \end{bmatrix}, t = 1, 2, \ldots$$

and the observation equation:

$$Y_t = \left[G(-1)^{d+1}\binom{d}{d} (-1)^d \binom{d}{d-1} (-1)^{d-1}\binom{d}{d-2} \ldots d \right] \begin{bmatrix} X_t \\ Y_{t-1} \end{bmatrix}, t = 1, 2, \ldots$$

with initial conditions $T_1 = \begin{bmatrix} X_1 \\ Y_0 \end{bmatrix} = \left[\sum_{j=0}^{\infty} F^j V_{-j} \right]$

and the assumption $E(Y_0 z'_t)$, $t = 0, \pm 1,...$

where $Y_0 = (Y_{1-d}, Y_{2-d},..., Y_0)'$.

Here, Y_0 is considered to be non-random in order to satisfy the state-space model and this implies $E(X_1 Y_0) = 0$ and $E(Y_0 \nabla^d Y'_t) = 0$, $t \geq 1$ as required for forecasting in ARIMA models.

This technique can be extended to SARIMA(P, D, Q) models also.

Example 10   *ARIMA(1,1,1) Model*

ARIMA(1,1,1) model is defined as

$$(1 - \phi B)(1 - B) Y_t = (1 + \theta B)z_t, \; \{z_t\} \sim \text{WN}(0, \sigma^2)$$

The state-space representation is given by

$$Y_t = [0 \; 1 \; 1] \begin{bmatrix} X_{t-1} \\ X_t \\ Y_{t-1} \end{bmatrix}$$

where $\begin{bmatrix} X_t \\ X_{t-1} \\ Y_1 \end{bmatrix} = \begin{bmatrix} 0 & 1 & 0 \\ 0 & \phi & 0 \\ 0 & 1 & 1 \end{bmatrix} \begin{bmatrix} X_{t-1} \\ X_t \\ Y_{t-1} \end{bmatrix} + \begin{bmatrix} 0 \\ z_{t+1} \\ 0 \end{bmatrix}$, $t = 1, 2,...$

and $[X_0, X_1, Y_0]' = \left[\sum_{j=0}^{\infty} \phi^j z_{-j}, \sum_{j=0}^{\infty} \phi^j z_{1-j}, Y_0 \right]$

5.2.4 Kalman Recursives

Kalman recursives are very powerful methods in solving three different problems concerned with finding the minimum MSE linear estimates of state-vector X_t in terms of observations $Y_1, Y_2,...$ and a random vector Y_0 which is orthogonal to V_t and W_t for all $t \geq 1$. In many cases, Y_0 is the constant vector $(1,1,...,1)'$ and the three problems for estimation of X_t are:

(i) : $Y_0,.... Y_{t-1}$ defines the prediction problem;
(ii) : $Y_0,...., Y_t$ defines the filtering problem; and
(iii) : $Y_0, ..., Y_n$ ($n > t$) defines the smoothing problem.

Definition

The best linear predictor of X_i in terms of all components $Y_0, Y_1,..., Y_t$ is defined at $P_t(X_i) = P(X \mid Y_0, Y_1,..., Y_t)$ hence, $P_t(X) = (P_t(X_1), ..., P_t(X_v))'$ is the best linear predictor for any random vector $X = (X_1, ..., X_v)'$.

The predictor is a linear component of observation values $Y_0, Y_1,..., Y_t$ such that $[X - P_t(X)]$ is orthogonal to Y_s, $s = 0,..., t$. If X, Y_0, $Y_1,..., Y_t$ are jointly normal and $Y_0 = (1,..., 1)'$, then $P_t(X) = E(X \mid Y_1,..., Y_t)$, $t \geq 1$. P_t is linear, i.e. $P_t(AX) = AP_t(X)$ and $P_t(x + v) = P_t(x) + P_t(v)$.

If X and Y are random vectors with v and w components and each having finite second moments, then

$$P(X \mid Y) = MY$$

where M is $v \times w$ matrix $= E(XY') [E(YY')]^{-1}$;

where $[E(YY')]^{-1}$ is any generalized inverse of matrix $E(YY')$. This prediction, fitting and smoothing problems reduce to determination $P_{t-1}(X_t)$, $P_t(X_t)$ and $P_n(X_t)$ $(n > t)$, respectively.

(i) Kalman prediction

For the state-space model, the one-step predictors $\hat{X}_t = P_{t-1}(X_t)$ and their error covariance matrices $\Omega_t = E[(X_t - \hat{X}_t) (X_t - \hat{X}_t)']$ and recursively for $t = 1, 2...$

$$\hat{X}_{t+1} = F_t \hat{X}_t + \Theta_t \Delta_t^{-1}(Y_t - \hat{X}_t)$$

$$\Omega_{t+1} = F_t \Omega_t F_t' + Q_t - \Theta_t \Delta_t^{-1} \Theta_t'$$

where $\Delta t = G_t \Omega_t G_t' + R_t$; $\Theta_t = F_t \Omega_t G_t'$ and Δ_t^{-1} is generalized inverse of Δ_t.

We can now use this for obtaining h-step best linear MSE predictors $P_t Y_{t+h}$; $h = 1, 2, ...$

We know,

$$P_t X_{t+1} = F_t P_{t-1} X_t + \Theta_t \Delta_t^{-1} (Y_t - P_{t-1} Y_t)$$

$$P_t X_{t+h} = F_{t+h-1} P_t X_{t+h-1}$$

$$= \cdots = (F_{t+h-1} F_{t+h-2} \cdots F_{t+1}) P_t X_{t+1}; h = 2, 3, ...$$

$$P_t Y_{t+h} = G_{t+h} P_t X_{t+h}, h = 1, 2, ...$$

We also get $\Omega_t^{(h)} = F_{t+h-1} \Omega_t^{(h-1)} F_{t+h+1}' + Q_{t+h-1}; h = 2, 3, ...$

with $\Omega_t^{(1)} = \Omega_{t+1}$, we also get

$$\Delta_t^{(h)} = G_{t+h} \Omega_t^{(h)} G_{t+h}' + R_{t+h}; h = 1, 2, ...$$

Example 11   *Random walk*

$$Y_t = X_t + W_t, \{W_t\} \sim WN(0, \sigma_v^2)$$

where local level X_t follows the random walk

$$X_{t+1} = X_t + V_t, \{V_t\} \sim WN(0, \sigma_w^2)$$

using the Kalman prediction equations with $Y_0 = 1$, $R = \sigma_w^2$ and $Q = \sigma_v^2$, we obtain

$$\hat{Y}_{t+1} = P_t Y_{t+1} = \hat{X}_t + \frac{\Theta_t}{\Delta_t}(Y_t - \hat{Y}_t)$$

$$= (1 - Q_t)\hat{Y} + a_t Y_t$$

where $a_t = (\Theta_t / \Delta_t) = \Omega_t / (\Omega_t + \sigma_w^2)$.

For a state-space model (like this one) with time-independent parameters, the solution of the Kalman recursives is called a steady-state solution if Ω_t is independent of t. If $\Omega_t = 2$ for all t, then we have

$$\Omega_{t+1} = \Omega = \Omega + \sigma_v^2 - (\Omega^2/(\Omega + \sigma_w^2))$$

$$= (\Omega \sigma_w^2 / (\Omega + \sigma_w^2)) + \sigma_v^2$$

Solving this quadratic equation for Ω and noting that $\Omega \geq 0$, we find

$$\Omega = (\sigma_v^2 + \sqrt{\sigma_v^2 + 4\sigma_v^2 \sigma_w^2})/2.$$

Regardless of value of Ω_1, Ω_t converges to Ω, the unique solution of $\Omega_{t+1} = \Omega_t$.

For any initial predictions $\hat{Y}_1 = \hat{X}_1$ and any initial mean square error, $\Omega_1 = E(X_1 - \hat{X}_1)^2$, the coefficients $a_t = \Omega_t / (\Omega_t + \sigma_w^2)$ converge to $a = \Omega/(\Omega + \sigma_w^2)$ and the MSE of predictors defined by $\hat{Y}_{t+1} = (1 - a_t)\hat{Y}_t + a_t Y_t$ converge to $\Omega + \sigma_w^2$. If Ω_1 is unknown, we cannot obtain $\{a_t\}$, so we consider

$$\hat{Y}_{t+1} = (1 - a)\hat{Y}_t + a Y_t$$

with $a \equiv a_t$ and arbitrary $\hat{Y}_1$.

Then, the sequence of predictions is also asymptotically optimal in MSE converging to $\Omega + \sigma_w^2$ as $t \to \infty$.

For an MA(1), we have $D_t = Y_t - Y_{t-1}$ as

$$D_t = z_t + \theta z_{t-1}, \{z_t\} \sim WN(0, \sigma^2)$$

where $\theta/(1 + \theta^2) = -\sigma_w^2/(2\sigma_w^2 + \sigma_v^2)$.

We get $\theta = -(2\sigma_w^2 + \sigma_v^2 - \sqrt{\sigma_v^2 + 4\sigma_v^2\sigma_w^2})/(2\sigma_w^2)$ and that $\theta = a - 1$.

We can derive an exponential smoothing formula for $\hat{Y}_t$ directly from ARIMA(0, 1, 1) structure of $\{Y_t\}$. For $t \geq 2$, we get

$$\hat{Y}_{t+1} = Y_t + \theta_{t1}(Y_t - \hat{Y}_t) = -\theta_{t1}\hat{Y}_t + (1 + \theta_{t1})Y_t$$

where θ_{t1} is found by innovations algorithm.

So, $1 - a_t = -\theta_{t1}$ and since $\theta_{t1} \to \theta$ and a_t converges to a steady state solution a, we conclude that $1 - a = \lim_{t \to \infty}(1 - a) = -\lim_{t \to \infty} \theta_{t1} = -\theta$.

(ii) Kalman filtering

The filtered estimates $X_{t|t} = P_t X_t$ and their error covariance matrices

$$\Omega_{t|t} = E[(X_t - X_{t|t})(X_t - X_{t|t})']$$

are defined by relating $P_t X_t = P_{t-1} X_t + \Omega_t G_t' \Delta_t^{-1}(Y_t - G_t \hat{X}_t)$ and $\Omega_{t|t} = \Omega_t - \Omega_t G_t' \Delta_t^{-1} G_t \Omega_t'$.

Proof:

$$P_t X_t = P_{t-1} X_t + MI_t$$

where $\quad M = E(X_t I_t')[E(I_t I_t')]^{-1}$

$$= E[X_t(G_t(X_t - \hat{X}_t) + W_t)']\Delta_t^{-1} = \Omega_t G_t' \Delta_t^{-1}$$

To find $\Omega_{t|t}$, we write

$$X_t - P_{t-1} X_t = X_t - P_t X_t + P_t X_t - P_{t-1} X_t = X_t - P_t X_t + MI_t$$

Since $X_t - P_t X_t$ and MI_t are orthogonal, we find $\Omega_t = \Omega_{t|t} + \Omega_t G_t' \Delta_t^{-1} \Omega_t'$ as required.

(iii) **Kalman fixed-point smoothing**

The smoothed estimates $X_{t|n} = P_n X_t$ and error matrices $\Omega_{t|n} = E[(X_t - X_{t|n}) \times (X_t - X_{t|n})']$ are defined for fixed t by the following recursives, which are solved successively for $n = t, t + 1, \ldots$

$$P_n X_t = P_{n-1} X_t + \Omega_{t,n} \, G_t' \Delta_t^{-1} (Y_n - G_n \hat{X}_n)$$

$$\Omega_{t,n+1} = \Omega_{t,n} [F_n - \Theta_n \Delta_t^{-1} G_n]'$$

$$\Omega_{t|n} = \Omega_{t|n-1} - \Omega_{t,n} G_t' \Delta_t^{-1} G_n \Omega_{t,n}'$$

with initial conditions $P_n X_t = \hat{X}_t$ and $\Omega_{t,t} = \Omega_{t|t-1} = \Omega_t$ (found from Kalman predictions).

We can estimate the parameters of structural parameters of state-space models through MLE, using the Kalman prediction recursives as follows:

1. For fixed Q^*, apply Kalman prediction recursives with $\hat{X}^* = 0$, $\Omega_1 = 0$, $Q = Q^*$ and $r_w^2 = 1$ to obtain predictors $\hat{X}_t^*$. Let Δ_t^* be one-step prediction errors for the above recursives.

2. Set $\hat{\mu} = \hat{\mu}(Q^*) = \left[\sum_{t=1}^{n} C_t' G' G C_t / \Delta_t \right]^{-1} \sum_{t+1}^{n} C_t' G' (Y_t - \hat{X}_t^*) / \Delta_t^*$.

3. Let $\hat{Q}^*$ be minimizer of $l(\hat{Q}^*)$, likelihood function

4. MLE of μ, Q, σ_w^2 are $\hat{\mu}, \hat{Q}^*, \hat{\sigma}_w^2$, where $\hat{\mu}, \hat{\sigma}_w^2$ are evaluated at $\hat{Q}^*$.

5.2.5 Missing Observations

State-space models with associated Kalman recursives are ideally suited for data analysis in such cases when a few observations may be missing and can be easily estimated. One method is to use the Gaussian likelihood based on $\{Y_{i1}, \ldots, Y_{ir}\}$, where i_j's are positive integers in such a manner that $1 \leq i_j \leq i_r \leq n$. This allows us to observe the process $\{Y_t\}$ at irregular intervals, i.e. $(n - r)$ observations are missing for sequence $\{Y_1, \ldots, Y_n\}$. If the solution is found, then we can obtain ARMA and ARIMA processes with missing values. A second method is the consideration of minimum MSE for the missing values themselves.

(i) **Gaussian likelihood**

The likelihood in this model with no missing observations for the state-space representation is

$$L(\theta, Y_1, ..., Y_n) = (2\pi)^{-nw/2} \left(\prod_{j=1}^{n} \det \Delta_j \right)^{-1/2} \exp\left[-\frac{1}{2} \sum_{j=1}^{n} I_j' \Delta_j^{-1} I_j \right]$$

where $I_j = Y_j - P_{j-1}Y_j$ and Δ_j, $j \geq 1$ are one-step predictors and error covariance matrices with $Y_0 = 1$.

For irregularly-spaced observations, we denote $\{Y_t^*\}$ related to the process $\{X_t\}$ by modified observation equation

$$Y_t^* = G_t^* X_t + W_t^*, t = 1, 2, ...$$

where $G_t^* = \begin{cases} G_t & \text{if } t \in \{i_1, ..., i_r\} \\ 0 & \text{otherwise} \end{cases}$ and $W_t^* = \begin{cases} W_t & \text{if } t \in \{i_1, ..., i_r\} \\ N_t & \text{otherwise} \end{cases}$

and $\{N_t\}$ is iid with $N_t \sim N(0, I_{w \times w})$, $N_s \perp X_1$,

$$N_s \perp \begin{bmatrix} V_t \\ W_t \end{bmatrix}, s, t = 0, \pm 1, ...$$

We have

$$L_1(\theta; y_{i1}, ..., y_{i_r}) = (2\pi)^{(n-r)w/2} L_2(\theta; y_1^*, ..., y_n^*)$$

where L_2 is the Gaussian likelihood for observed data.

Then, $L_1(\theta; y_{i}, ..., y_{i_r}) = (2\pi)^{-rw/2} \left(\prod_{j=1}^{n} \det \Delta_j^* \right)^{1/2} \exp\left[-\frac{1}{2} \sum_{j=1}^{n} i_j^{*'} \Delta_j^{*-1} i_j^* \right]$

where i_j^* denotes the observed innovation $y_j^* - \hat{y}_j^*$; $j = 1, 2, ..., n$.

Example 12 *One missing observation in AR(1) series*
Let $\{Y_t\}$ be the causal AR(1) process $Y_t - \phi Y_{t-1} = z_t$, $\{z_t\} \sim WN(0, \sigma^2)$.
Suppose we have five data $y_1, y_2, ..., y_5$ of which y_2 is missing. Then, $y_i^* = y_i$, $i = 1, 3, 4, 5$ and $y_2^* = 0$. The state-space model for $\{Y_t\}$ is $Y_t = X_t$, $X_{t+1} = \phi X_t + z_{t+1}$.

The corresponding model for $\{y_t^*\}$ is then

$$y_t^* = G_t^* X_t + W_t^*, t = 1, 2, ...$$

where $X_{t+1} = F_t X_t + V_t, t = 1, 2, ...$

$$F_t = \phi, = G_t^* = \begin{cases} 1 & \text{if } t \neq 2 \\ 0 & \text{if } t = 2 \end{cases}, \quad V_t = z_{t+1}, \quad W_t^* = \begin{cases} 0 & \text{if } t \neq 2 \\ N_t & \text{if } t = 2 \end{cases}$$

$$Q_t = \sigma^2, \quad R_t^* = \begin{cases} 0 & \text{if } t \neq 2 \\ 1 & \text{if } t = 2 \end{cases}, \quad S_t^* = 0$$

and
$$X_1 = \sum_{j=0}^{\infty} \phi^j z_{1-j}.$$

Starting from the initial conditions $\hat{X}_1 = 0$, $\Omega_1 = \sigma^2/(1 - \phi^2)$ and applying the Kalman recursives, we find

$$\theta_t \Delta_t^{-1} = \begin{cases} \phi & \text{if } t \neq 2 \\ 0 & \text{if } t = 2 \end{cases}; \quad \Omega_t = \begin{cases} \sigma^2/(1 - \phi^2) & \text{if } t = 1 \\ \sigma^2(1 + \phi^2) & \text{if } t = 3 \\ \sigma^2 & \text{if } t = 2, 4, 5 \end{cases}$$

and $\hat{X}_1 = 0$, $\hat{X}_2 = \phi Y_1$, $\hat{X}_3 = \phi^2 Y_1$, $\hat{X}_4 = \phi Y_3$, $\hat{X}_5 = \phi Y_4$.

For $h = 1$, we find $\hat{Y}_1^* = 0$, $\hat{Y}_2^* = 0$, $\hat{Y}_3^* = \phi^2 Y_1$, $\hat{Y}_4^* = \phi Y_3$, $\hat{Y}_5^* = \phi Y_4$ with mean squared errors as

$$\Delta_1^* = \sigma^2/(1 - \phi^2), \quad \Delta_2^* = 1, \quad \Delta_3^* \sigma^2 (1 + \phi^2), = \sigma^2, = \sigma^2.$$

So, $L_1(\phi, \sigma^2, y_1, y_3, y_4, y_5) = \sigma^{-4} (2\pi)^{-2} [(1 - \phi^2) (1 + \phi^2)]^{1/2}$

$$\times \exp \left\{ -\frac{1}{2\sigma^2} \left[y_1^2 (1 - \phi^2) + \frac{(y_3 - \phi^2 y_1)^2}{(1 + \phi^2)} + (y_4 - \phi y_3)^2 + (y_5 - \phi y_4)^2 \right] \right\}.$$

This analysis can be extended to ARIMA (p, d, q) where the missing values among first d observations $y_{1-d}, \ldots, y_0$ can be handled as if they were unknown parameters for likelihood maximization.

(ii) Missing values for state-space models

We wish to find the minimum mean square error estimates $P(Y_t | Y_0, Y_{L1}, \ldots, Y_{lr})$ of Y_t, $1 \leq t \leq n$, where $Y_0 = 1$.

We use the modified process $\{Y_t^*\}$ with $Y_0^* = 1$.

Since $Y_s^* = Y_s$ for $s \in \{i_1, \ldots, i_r\}$ and $Y_s^* \perp X_t$, Y_0 for $1 \leq t \leq n$ and $s \notin \{0, i_1, \ldots, i_r\}$, we obtain minimum MSE state estimates

$$P(X_t \mid Y_0, Y_{i_1}, ..., Y_{i_r}) = P(X_t \mid Y_0^*, Y_1^*, ..., Y_n^*) , 1 \le t \le n.$$

The RHS is evaluated by applying the Kalman fixed-path smoothing algorithm to state-space model equations. For computation Y_t^*, $t \notin \{0, i_1, ..., i_r\}$ are immaterial and set equal to zero.

To evaluate $P(Y_t \mid Y_0, Y_{i_1}, ..., Y_{i_r})$, $1 \le t \le n$ we use relation $Y_t = G_t X_t + W_t$

Since $E(V_t W_t') = S_t = 0, t = 1, 2, ..., n$, we obtain

$$P(Y_t \mid Y_0, Y_{i_1}, ..., Y_{i_r}) = G_t P(X_t \mid Y_0^*, Y_1^*, ..., Y_n^*)$$

Example 13 *AR(1) series with one missing value*

Y_2 is the missing value and $Y_0 = 1$, and Y_1, Y_3, Y_4, Y_5 are observed. We have state-space model $X_{t+1} = \phi X_t + z_{t+1}, Y_t = X_t$ for $\{Y_t\}$.

The corresponding model for $\{Y_t^*\}$ is given in a previous example. Applying the Kalman smoothing equations, we find

$$P_1 X_2 = \phi Y_1; P_2 X_2 = \phi Y_1; P_3 X_2 = \phi(Y_1 + Y_3)/(1 + \phi^2)$$
$$P_4 X_2 = P_3 X_2; P_5 X_2 = P_3 X_2$$
$$\Omega_{2,2} = \sigma^2; \Omega_{2,3} = \phi \sigma^2; \Omega_{2,t} = 0, t \ge 4.$$

and
$$\Omega_{2 \mid t} = \sigma^2; \Omega_{2 \mid 2} = \sigma^2; \Omega_{2 \mid t} = \sigma^2/(1 + \phi^2), t \ge 3.$$

Above, $P_t(\cdot)$ denotes $P(\cdot \mid Y_0^*, Y_1^*, ..., Y_n^*)$ and $\Omega_{t,n}, \Omega_{t \mid n}$ are correspondingly defined.

The minimum MSE for Y_2 is found as

$$P_5 Y_2 = P_5 X_2 = \phi(Y_1 + Y_3)/(1 + \phi^2)$$

with MSE $\Omega_{2 \mid 5} = \sigma^2/(1 + \phi^2)$.

This can be easily extended to the missing values for ARIMA(p, d, q) process.

In addition, Kalman recursive approaches are rather involved, so we can use the direct approach of maximizing the joint distribution of x, y when both are MND and a few observations are missing.

We write $\qquad f_{x,y}(x, y) = f_{x \mid y}(x \mid y) f_y(y)$

where $f_{x \mid y}(x \mid y)$ is also multivariate normal with mean $E(X \mid Y)$ and covariance matrix $\Sigma_{x \mid y}$. We have

$$f_{x,y}(x|y) = \frac{1}{\sqrt{(2\pi)^q \det \Sigma_{x|y}}} \exp\left\{-\frac{1}{2}(x - E(X|y))' \sum\nolimits_{x|y}^{-1} (x - E(X|y))\right\}$$

where $q = \dim(X)$.

The maximum is reached when $x = E(X|y)$, so the best estimator of X in terms of Y can be found by maximizing the joint density of X and Y with respect to x. For AR model, it is easy to carry out this optimization.

Example 14 *Missing values in AR process*

Suppose $\{Y_t\}$ is AR(p) process $Y_t = \phi_1 Y_{t-1} + \cdots + \phi_p Y_{t-p} + z_t$, $\{z_t\} \sim$ WN $(0, \sigma^2)$ and $Y = (Y_{i_1}, ..., Y_{i_r})'$ with $1 \leq i_1 < \cdots < i_r \leq n$ are the observed values. If there are no missing values in first p observations, then both the best estimates

of the missing values are found by minimizing $\displaystyle\sum_{t=p+1}^{n} (Y_t - \phi_1 Y_{t-1} - \cdots -$

$\phi_p Y_{t-p})^2$ with respect to the missing values. For the AR(1) model with Y_2 missing, this is same as minimizing

$$(Y_2 - \phi_1 Y_1)^2 + (Y_3 - \phi_1 Y_2)^2$$

with respect to Y_2.

So, differentiating with respect to Y_2 and setting it equal to 0, we obtain

$$E(Y_2 | Y_1, Y_3, Y_4, Y_5) = \phi(Y_1 + Y_3)/(1 + \phi^2)$$

5.2.6 EM Algorithm

The expectation-maximization (EM) algorithm is an iterative scheme to compute MLE only when a subset of complete data set is available and hence it can be used for estimation of missing observations as well. Complete data vector w is made up of 'observed' data Y and 'unobserved' data X (or state vectors $X_1, ..., X_n$). Each iteration of the EM algorithm has two steps. If $\theta^{(i)}$ denotes the estimated value of parameter θ after i iterates, then the two steps in the $(i + 1)$ iteration are:

E-step: Calculate $Q(\theta|\theta^{(i)}) = E_{\theta^{(i)}}[l(\theta_j, X, Y)|Y]$

M-step: Maximize $Q(\theta|\theta^{(i)})$ with respect to θ.

$\theta^{(i+1)}$ is then set equal to the maximizer Q in the M-step.

In the E-step, $l(\theta : x, y) = \ln f(x, y; \theta)$ and $E_{\theta(i)}(\cdot|Y)$ denotes conditional expectation relative to the conditional density $f(x|y, \theta^{(i)}) = f(x, y, \theta^{(i)})/f(y, \theta^{(i)})$. Since $l(\theta^{(i)}; Y)$ is non-decreasing in i, $\theta^{(i)}$ has a limit $\hat{\theta}$ which is the solution of the likelihood equation $l'(\hat{\theta}; Y) = 0$. The advantage of EM algorithm over direct maximization of likelihood is that the former (for maximization of Q) is often much easier in the M-step.

Missing Data: Assume $W = (X', Y)'$ has a multivariate normal distribution with mean 0 and covariance matrix Σ, which depends on parameters θ. The log likelihood for complete data is given by

$$l(\theta; W) = -\frac{n}{2} \ln(2\pi) - \frac{1}{2} \ln \det(\Sigma) - \frac{1}{2} W'\Sigma W$$

The E-step requires that we compute expectation of $l(\theta; W)$ with respect to conditional distribution of W given Y with $\theta = \theta^{(i)}$. Writing $\Sigma(\theta)$ as the block matrix

$$\Sigma = \begin{bmatrix} \Sigma_{11} & \Sigma_{12} \\ \Sigma_{21} & \Sigma_{22} \end{bmatrix}$$

which is conformable with X and Y the conditional distribution of W given Y is multivariate normal with mean $\begin{bmatrix} \hat{x} \\ 0 \end{bmatrix}$ and covariance matrix

$\begin{bmatrix} \Sigma_{1|2}(\theta) & 0 \\ 0 & 0 \end{bmatrix}$, where $\hat{X} = E_\theta(X|Y) = \Sigma_{12}\Sigma_{22}^{-1}Y$ and $\Sigma_{1|2}(\theta) = \Sigma_{11} - \Sigma_{12}\Sigma_{22}^{-1}\Sigma_{21}$.

We then have $E_{\theta^{(i)}}[(X', Y')\Sigma(X' Y')' | Y] = \text{trace}(\Sigma_{1|2}(\theta^{(i)})\Sigma_{1|2}^{-1}(\theta)) + \hat{W}'\Sigma^{-1}(\theta)\hat{W}$, where $\hat{W} = (\hat{X}', Y)'$.

It follows that $Q(\theta| \theta^{(i)}) = l(\theta, \hat{W}) - \frac{1}{2}\text{trace}(\Sigma_{1|2}(\theta^{(i)})\Sigma_{1|2}^{-1}(\theta))$.

If the increments $\theta^{(i+1)} - \theta^{(i)}$ are small, then the second term on RHS of $Q(\theta| \theta^{(i)})$ is nearly constant ($\approx n - r$) and can be ignored. So, we use

$$\tilde{Q}(\theta|\theta^{(i)}) = l(\theta; \hat{W})$$

Then EM algorithm is as follows:

E-step: Calculate $E_{\theta^{(i)}}(X|Y)$ (e.g. with Kalman fixed point smoothing and form $l(\theta; \hat{W})$)

M-step: Find the MLE for the "complete" data problems, i.e. maximize $l(\theta; \hat{W})$ using the best estimates computed in *E-step* for the missing values.

We can also generalize state-space models to a multi-dimensional state and observation variables without much change, i.e. $X_t \rightarrow \mathbf{X}_t$ and $Y_t \rightarrow \mathbf{Y}_t$. This could be either parameter-driven or observation-driven.

5.3   DYNAMIC LINEAR MODELLING (DLM) WITH SWITCHING

Often there occur regime changes for vector-valued time series. For example, due to dynamics of process change over a period of time, the response may be different during expansion ($\nabla \log y_t > 0$) than during contraction ($\nabla \log y_t < 0$). We assume that the dynamics of the underlying model changes discontinuously at certain unknown time points. We wish to find these time points and the dynamic relation prior and later to these time points.

Let DLM be given by $x_t = \Phi\, x_{t-1} + w_t$

This describes the $p \times 1$ state (dynamic) equation but we observe $y_t = A_t\, x_t + v_t$ to describe $q \times 1$ observation (dynamic) equation, where w_t and v_t are Gaussian white noise sequences with Var $(w_t) = Q$, Var $(v_t) = R$ and Cov $(w_t, v_s) = 0$ for all s and t.

Switching at time t may occur in a stationary Markov chain in one of the m possible regimes. Suppose the dynamics of a univariate time series y_t is generated by either the model (1) $y_t = \beta_1\, y_{t-1} + w_t$ or the model (2) $y_t = \beta_2\, y_{t-1} + w_1$. Then we write the model as $y_t = \phi_t\, y_{t-1} + w_t$ such that $Pr\,(\phi_t = B_j) = \pi_j; j = 1, 2; \pi_1 + \pi_2 = 1$ and with Markov property $Pr\,(\phi_t = B_j | \phi_{t-1} = B_i, \phi_{t-2} = B_{i2}) = Pr\,(\phi_t = B_j | \phi_{t-1} = B_i) = \pi_{ij}$ for $i, j = 1, 2$ and $(i_2, \ldots = 1, 2)$. This is also called hidden (unknown to observer) Markov model.

Example 15   *Economic change*

This is applicable for dynamics of linear model changes that occur suddenly over the history of a given realization (for example, positive/negative growths; high grade/low grade ore mineralization).

Suppose data is governed by

$$y_t = x_t + n_t$$

where x_t is an AR series and n_t is a random walk with drift that switches between two values α_0 and $\alpha_0 + \alpha_1$.

Thus, $n_t = n_{t-1} + \alpha_0 + \alpha_1 S_t$ with $S_t = 0$ or 1 if the system is in state 1 or state 2.

Let $x_t = \phi x_{t-1} + \phi_2 x_{t-2} + w_t$ be AR(2) with Var$(w_t) = \sigma_w^2$. Then, we have

$$\nabla y_t = x_t - x_{t-1} + \alpha_0 + \alpha_1 S_t$$

which is the observation equation with state vector

$$x_t = (x_t, x_{t-1}, \alpha_0, \alpha_1)'$$

and $M_1 = [1, -1, 1, 0]$ and $M_2 = [1, -1, 1, 1]$, determining two economic conditions. The state equation is of the form

$$\begin{pmatrix} x_t \\ x_{t-1} \\ \alpha_0 \\ \alpha_1 \end{pmatrix} = \begin{bmatrix} \phi_1 & \phi_2 & 0 & 0 \\ 1 & 0 & 0 & 0 \\ 0 & 0 & 1 & 0 \\ 0 & 0 & 0 & 1 \end{bmatrix} \begin{pmatrix} x_{t-1} \\ x_{t-2} \\ \alpha_0 \\ \alpha_1 \end{pmatrix} + \begin{pmatrix} w_t \\ 0 \\ 0 \\ 0 \end{pmatrix}$$

The observation equation in this case becomes

$$\nabla y_t = A_t x_t + v_{tj}$$

where $Pr(A_t = M_1) = 1 - Pr(A_t = M_2)$, with M_1, M_2 given as above.

We now assume that the m possible configurations are states in a non-stationary, independent process, as defined by the time-varying probabilities

$$\pi_j(t) = Pr(A_t = M_j)$$

for $j = 1, 2, ..., m$ and $t = 1, 2, ..., n$.

Independent information about the current state of measurement process is given by the filtered probabilities of being in state j, defined as conditional probabilities

$$\pi_j(t \mid t) = Pr(A_t = M_j \mid Y_t)$$

which also vary as a function of time.

We should find estimators for $\pi_j(t \mid t)$, predicted state x_t^{t-1} and predicted state x_t^t and corresponding error covariance matrices P_t^{t-1} and P_t^t. The parameters Θ of DLM are unknown and we estimate by MLE (but Bayesian assignment can also be made for Θ).

We have

$$\text{predictions } x_t^{t-1} = \Phi\, x_{t-1}^{t-1}; \; P_t^{t-1} = \phi P_{t-1}^{t-1} \phi' + Q$$

$$\text{filters } x_t^t = x_t^{t-1} + \sum_{j=1}^{m} \pi_j(t \mid t) \kappa_{tj} \varepsilon_{tj}$$

$$P_t^t = \sum_{j=1}^{m} \pi_j(t \mid t)(I - \kappa_{tj} M_j) P_t^{t-1}$$

$$\kappa_{tj} = P_t^{t-1} M'_j \Sigma_{tj}^{-1}$$

$$\varepsilon_{tj} = y_t - M_j x_t^{t-1}$$

$$\Sigma_{tj}^{-1} = M_j P_t^{t-1} M'_j + R$$

for $j = 1, ..., m$.

Next, we derive $\pi_j(t|t)$ as

$$\pi_j(t|t) = \frac{\pi_j(t)f_j(t|t-1)}{\sum_k^m \pi_k(t)f_k(t|t-1)}$$

where $f_j(t|t-1)$ denotes the conditional density of y_t given past $y_1, y_2, ...,$ y_{t-1} and $A_t = M_j$ for $j = 1, ..., m$. If any specific state is not preferred, then uniform priors, $\pi_j(t) = m^{-1}$, for $j = 1, ..., m^{-1}$ for $j = 1, ..., m$ will be applied. Smoothness is obtained by letting

$$\text{(smoothed)} \ \pi_j(t) = \sum_{i=1}^m \pi_i(t-1|t-1)\pi_{ij}$$

where π_{ij} are non-negative weights such that $\sum_{i=1}^m \pi_{ij} = 1$.

If A_t process was Markov with transition probabilities π_{ij}, then $\pi_j(t)$ would update the filter probability. However, these techniques are highly computer intensive.

Example 16 *Ore grades in ore-shoot/non-ore-shoot zones*

The ore grade suddenly increases to a considerably higher level in the ore-shoots, whereas in other areas of mineralization, it may be mineable but not of very high grades. The positions of ore grades are not exactly known prior to actual mining of the ores. Therefore, on-line recognition of ore-shoots would be useful for storage, blending, beneficiation and marketing purposes. The variations in the assay values within ore-shoots and non-ore-shoots may be modelled as DLM with growth and contraction, respectively. We may have two components.

(1) a fixed trend $x_{t1} = \alpha x_{t-1,1} + w_{t1}$, where w_{t1} is white noise with Var$(w_{t1}) = \sigma_1^2$; and

(2) a second component x_{t2}, representing the sharp rise within ore-shoots given by

$$x_{t2} = \beta_0 + \beta_1 x_{t-1,2} + w_{t2}$$

where w_{t2} is white noise with Var$(w_{t2}) = \sigma_2^2$.

In the non-ore-shoot (state 1) we have

$$y_t = x_{t1} + v_t$$

where v_t is measurement errors which is white noise with $\text{Var}(v_t) = \sigma_v^2$. In ore-shoots, we have

$$y_t = x_{t1} + x_{t2} + v_t.$$

In the state-space form, this can be written as the state equation

$$\begin{pmatrix} x_{t1} \\ x_{t2} \end{pmatrix} = (\alpha_1) \begin{pmatrix} x_{t-1,1} \\ x_{t-1,2} \end{pmatrix} + \begin{pmatrix} 0 \\ \beta_0 \end{pmatrix} + \begin{pmatrix} w_{t1} \\ w_{t2} \end{pmatrix}$$

or as $x_t = \Phi x_{t-1} + \alpha + w_t$
where $x_t = (x_{t1}, x_{t2})'$; $\alpha = (0, \beta_0)$ and Q is a 2×2 matrix with σ_1^2 as $(1,1)$ and σ_2^2 as $(2,2)$ elements.
The filter equation is given by

$$x_t^{t-1} = \Phi\, x_{t-1}^{t-1} + \alpha$$

The observation equation is

$$y_t = A_t x_t + v_t$$

where A_t is a 1×2 and v_t is white noise with x_t^{t-1} $\text{Var}(v_t) = R = \sigma_w^2$.
We assume w_{t1}, w_{t2}, v_1 are uncorrelated.
A_t can take two forms:

$$M_1 = [1\ 0] \text{ no ore-shoot}$$

$$M_2 = [0\ 1] \text{ ore-shoot}$$

such that $Pr(A_t = M_1) = 1 - Pr(A_t = M_2)$.

We again assume that A_t is the hidden Markov chain as location of ore-shoot is unknown and transition probabilities $\pi_{11} = \pi_{22} = 0.8$ (thus, $\pi_{12} = \pi_{21} = 0.2$).

We assume the initial values of $\pi_1(1\,|\,0) = 0.5 = \pi_2(1\,|\,0) = 0.5$ and estimate the different parameters of this model using non-linear maximization of the likelihood expression based on full data.

5.4 NON-LINEAR MODELS

5.4.1 Introduction

We use linear process for $\{X_t\}$ in a mere restrictive sense, i.e. $\{z_t\} \sim$ IID $(0, \sigma^2)$. A time series of the form

$$X_t = \sum_{j=0}^{\infty} \psi_j z_{t-j}, \quad \{z_t\} \sim \text{IID } (0, \sigma^2).$$

where z_t is expressible as a mean square limit of linear components of $\{X_s, \infty < s \leq t\}$ has the properties that the best mean square predictor $E(X_{t+h} \mid X_s, -\infty < s \leq t)$ and the best predictor $\tilde{P} X_{t+h}$ in terms of $\{X_s, -\infty < s \leq t\}$ are the same. If $\{z_t\}$ is not IID but WN sequence, then it is true if $E(z_t \mid X_s, -\infty < s \leq t) = 0$ for all t. As per the Wold decomposition, every purely non-deterministic stationary process can be expressed as above but with $\{z_t\} \sim$ IID $(0, \sigma^2)$. But WN is not necessarily an IID sequence and hence, the best mean square predictor may be quite different for the best linear predictor, based on $\{z\} \sim$ IID $(0, \sigma^2)$.

If $\{z_t\}$ is a purely non-deterministic Gaussian stationary process, the sequence $\{z_t\}$ in the Wold decomposition is Gaussian and, therefore, IID. Every stationary purely non-deterministic Gaussian process can, therefore, be generated by applying a causal linear filter to an IID Gaussian sequence. Such a process can be termed as Gaussian linear process.

Many time series have characteristics which are not valid for linear processes, i.e. minimum mean square predictors are not in general linear functions of past observations, but are non-linear.

Gaussian linear processes have characteristic time-reversibility or symmetry, but the observed time series may be asymmetric. For example, stress and strain build-ups before an earthquake may be gradual over a larger period of time, where the strain release during and after an earthquake is usually very fast, growth and decline of sunspots, sudden bursts in outlying values may give infinite variance (non-linear) ecological limit cycles (prey-predator model), chemical limit cycles in ore mineral deposition for mineralized solutions in ocean basins, etc. In many financial time series, we have periods of good and bad predictability which may be due to non-linearity, since linear predictors are independent of time history. In a mine, we may have poor predictability of ore grade within ore-shoots, whereas in other parts of mineralization, the predictability may be very good and these may be modelled by ARCH model, where CH stands for *conditional heteroscedasticity*.

In order to distinguish between the linear and non-linear processes, we have to decide whether a white noise sequence is also an IID sequence. Sequences generated by non-linear deterministic equations (for example, equation $x_n = 4x_{n-1} (1 - x_{n-1})$, $0 < x_0 < 1$) can generate sequences that are extremely close to white noise. Therefore, in order to distinguish between IID and non-IID white noise, we require moments of the data higher than second-order for testing stationarity of the series.

If $\{X_t\} \sim WN(0, \sigma^2)$ and $E \mid X_t \mid 4 < \infty$, a useful tool to decide whether $\{X_t\}$ is IID or not is by computing acf ρ_{X^2} (h) for the process $\{X_t^2\}$. If $\{X_t\}$ is IID, ρ_{X^2} (h) = 0 for all $h \neq 0$, otherwise not.

The third order cumulant function C_3 of $\{X_t\}$ with third order central moment function if $\{X_t\}$ is strictly stationary and $C_3(r, s) = E[(X_t - \mu) \times (X_{t+r} - \mu) (X_{t+s} - \mu)]$, $r, s \in \{0, \pm 1\}$, where $\mu = EX_t$.

If $\Sigma_r \Sigma_s \mid C_3$ (r, s) < ∞, third order polyspectral density (or bispectral density) of $\{X_t\}$, is the Fourier transform

$$f_3(w_1, w_2) \;=\; \frac{1}{(2\pi)^2} \sum_{r=-\infty}^{\infty} \sum_{x=-\infty}^{\infty} C_3(r, s)e^{-irw_1 - isw_2}, \; -\pi \leq w_1, w_2 \leq \pi.$$

or $\quad C_3 \ (r, s) \;=\; \int\limits_{-\pi}^{\pi} \int\limits_{-\pi}^{\pi} e^{-irw_1 + isw_2} \, f(w_1, w_2) \, dw_1 \, dw_2$

If $\{X_t\}$ is a linear process with $E \mid z_t \mid 3 < \infty$, $E z_t^3 = \eta$ and $\sum_{j=0}^{\infty} \mid \psi_j \mid$ < ∞, then

$$C_3(r, s) \;=\; \eta \sum_{i=-\infty}^{\infty} \psi_i \, \psi_{i+r} \, \psi_{i+s}$$

(with $\psi_j = 0$ for $j < 0$), where ψ are the transfer functions. Hence, $\{X_t\}$ has bispectral density

$$f_3(w_1, w_2) \;=\; \frac{\eta}{4\pi^2} \, \psi(e^{i(w_1 + w_2)}) \, \psi(e^{-iw_1}) \, \psi(e^{-iw_2})$$

where $\psi(z) = \sum_{j=0}^{\infty} \psi_j z^j$.

Therefore, the spectral density of $\{X_t\}$ is

$$f(w) \;=\; \frac{\sigma^2}{2\pi} \mid \psi(e^{-iw}) \mid^2$$

Hence, $\phi(w_1, w_2) \;=\; \dfrac{\mid f_3(w_1, w_2) \mid^2}{f(w_1) \, f(w_2) \, f(w_1 + w_2)} \;=\; \dfrac{\eta^2}{2\pi\sigma^6}$

Therefore, a linear process is an appropriate model for $\{X_t\}$ provided $\phi(w_1, w_2)$ is constant $\left(= \dfrac{\eta^2}{2\pi\sigma^6} \right)$.

5.4.2 Main Classes of Non-linear Models

If we decide that linear time series models are not appropriate and useful for the observed series, then we have to choose one of the non-linear models which fit the data well and give good forecasts. There are several families of non-linear time series models:

   (i) Bilinear
  (ii) AR with random coefficients
 (iii) Exponential AR (EAR)
  (iv) Threshold AR
   (v) State-dependent models (SDM)

The bilinear model of order (p, q, r, s) is defined as

$$X_t = z_t + \sum_{i=1}^{p} a_i X_{t-i} + \sum_{j=1}^{q} b_j z_{t-j} + \sum_{i=1}^{r} \sum_{i=1}^{x} c_{ij} X_{t-i} z_{t-j}$$

$$|\leftarrow \quad \text{Linear ARMA} \quad \rightarrow| \quad |\leftarrow \quad \text{Nonlinear} \quad \rightarrow|$$

where $\{z_t\} \sim$ IID $(0, \sigma^2)$.

A random coefficient AR process $\{X_t\}$ of order p has the form

$$X_t = \sum_{i=1}^{p} (\phi_i + U_t^{(i)}) X_{t-i} + z_t$$

where $\{z_t\} \sim$ IID $(0, \sigma^2)$, $\{U_t^{(i)}\} \sim$ IID $(0, v^2)$.
$\{z_t\}$ is independent of $\{U_t\}$ and $\phi_1, ..., \phi_p \in R$.

Threshold models are piecewise linear, in which the linear relationship varies with values of the process. In an $AR_{(p)}$ model $R^{(i)}$, $i = 1, ..., k$ is a partition of R^p, $\{z_t\} \sim$ IID $(0, \sigma^2)$, then the k-difference equations are

$$X_t = \sigma^{(i)} z_t + \sum_{j=1}^{p} \phi_j^{(i)} X_{t-j}, (X_{t-1}, ..., X_{t-p}) \in R^{(i)}, i = 1,..., k.$$

Model identification and parameter estimation for threshold models are exactly similar as for linear models using MLE and AIC criterion.

Exponential autoregressive (EAR) models are useful for modelling random oscillations in time/spatial domains as in case of sedimentary cycles, climate cycles, etc. Here the coefficients change smoothly rather than discontinuously as in TAR models.

State-dependent models (SDM) are more general and include all the above models, but in a state-space framework. These are locally linear ARMA models with local mean, model parameters dependent on the state of process at time $(t-1)$ and hence can yield linear, bilinear, TAR, EAR models with suitable assumptions.

In order to solve the non-linear models, we also require transfer functions for a higher (say maximum up to 10) order moments (≥ 2) but may delete those transfer functions (order < 10) which have negligible values.

5.4.3 ARCH(p) Model

We define stock returns as $\nabla [\ln y_t]$, where y_t is the stock price. ARCH models are used to model volatility changes in stock returns (or variance changes in assays in ore-shoots versus mineralized zones) under deviations from linearity (refer Bollerslev, 1986; Tsay, 1987)

ARCH(p) process $\{X_t\}$ is a solution of the equations

$$X_t = \sigma_1 z_t; \{z_t\} \sim \text{IID } N(0, 1)$$

where σ_t is the positive function of $\{X_s, s < t\}$ defined by

$$\sigma_t^2 = \alpha_0 + \sum_{i=1}^{p} \alpha_i X_{t-1}^2$$

with $\alpha_0 > 0$ and $\alpha_j \geq 0$, $j = 1,..., p$.

ARCH represents *autoregressive conditional heteroscedasticity* of random variable.

The simplest process would be ARCH(1) process. Then, we obtain

$$X_t^2 = \alpha_0 z_t^2 + \alpha_1 X_{t-1}^2 z_t^2$$
$$= \alpha_0 z_t^2 + \alpha_1 \alpha_0 z^2 z_{t-1}^2 + \alpha_1^2 X_{t-2}^2 z_t^2 z_{t-1}^2$$

$$= \cdots = \alpha_0 \sum_{j=0}^{n} \alpha_1^j z_t^2 z_{t-1}^2 \ldots z_{t-j}^2 + \alpha_1^{n+1} X_{t-n+1}^2 z_t^2 z_{t-1}^2 \ldots z_{t-n}^2$$

If $|\alpha_1| < 1$ and $\{X_t\}$ is stationary and causal (i.e. X_t is a function of $\{z_s, s \leq t\}$), then the expectation of the last term converges to 0 and the expectation of the sum converges to $\alpha_0 \sum_{j=0} \alpha_1^j = \alpha_0/(1 - \alpha_1) < \infty$ as $n \to \infty$.

It follows that the series, $\alpha_0 \sum_{j=0}^{n} \alpha_1^j z_t^2 z_{t-1}^2 \ldots z_{t-j}^2$ converges and hence

$$X_t^2 = \alpha_0 \sum_{j=0}^{\infty} \alpha_1^j z_t^2 z_{t-1}^2 \ldots z_{t-j}^2$$

So, $E X_t^2 = \alpha_0/(1 - \alpha_1)$; $\text{Var}(X_t^2) = EX_t^4 = \dfrac{3\alpha_0^2}{(1-\alpha_1)^2} \dfrac{1-\alpha_1^2}{1-3\alpha_1^2}$ if $3\alpha_1^2 < 1$.

Thus X_t has fat tails compared to Normal, i.e. leptokurtic.

Since $X_t = z_t \sqrt{\alpha_0 \left(1 + \sum_{j=0}^{\infty} \alpha_1^j z_{t-1}^2 \dots z_{t-j}^2 \right)}$ it is clear that $\{X_t\}$ is
strictly stationary and hence $EX_t^2 < \infty$, also stationary in the weak sense.
Therefore, the solution of ARCH(1) equations are:
$|\alpha_1| < 1$, the unique causal stationary solution of the ARCH(1) equations
is given by

$$X_t = z_t \sqrt{\alpha_0 \left(1 + \sum_{j=0}^{\infty} \alpha_1^j z_{t-1}^2 \dots z_{t-j}^2 \right)}$$

It has following properties:

$$E(X_t) = E(E(X_t | z_s, s < t)) = 0; \; \mathrm{Var}\,(X_t) = \alpha_0/(1 - \alpha_1)$$

and $E(X_{t+h} X_t) = E(E(X_{t+h} X_t | z_s, s < t)) = 0 \text{ for } h > 0.$

Thus, the ARCH(1) process with $|\alpha_1| < 1$ is a strictly stationary white
noise but not an IID sequence, since $E(X_t^2 | X_{t-1}) = (\alpha_0 + \alpha_1 X_{t-1}^2)$ and
$E(z_t^2 | X_{t-1}) = \alpha_0 + \alpha_1 X_{t-1}^2$.

This shows that $\{X_t\}$ is not Gaussian, as strictly stationary Gaussian
noise is necessarily IID. The distribution of X_t is symmetric, i.e. X_t and X_{-t}
have the same distribution. We can unite the likelihood $2X_1$, ..., X_n
conditional $\{X_1,..., X_p\}$ and hence by numerical maximization, compute
conditional MLE of parameters. The conditional likelihood of
observations $\{x_2,..., x_n\}$ of the ARCH(1) process given $X_1 = x_1$ is

$$L = \prod_{t=2}^{n} \frac{1}{\sqrt{2\pi(\alpha_0 + \alpha_1 x_{t-1}^2)}} \exp\left\{ -\frac{x_t^2}{2(\alpha_0 + \alpha_1 x_{t-1}^2)} \right\}$$

Estimation of α_0, α_1 is done by minimizing $-2 \ln L (\alpha_0, \alpha_1 | x)$. A
generalization of ARCH(p) process is the GARCH(p) process in which the
variance is dependent on p, AR and q, MA parameters and we have

$$\sigma_t^2 = \alpha_0 + \sum_{i=1}^{p} \alpha_i X_{t-i}^2 + \sum_{j=1}^{q} \beta_j \sigma_{t-j}^2$$

with $\alpha_0 > 0$ and $\alpha_j, \beta_j \geq 0, j = 1, 2, \dots.$

Example 16   *US GNP Residuals*

The return rates showed a MA(2) fit from which the residuals are
computed. These residuals behaved as white noise process but many
authors suggested that GNP behaves as an ARCH process. Since the

residuals are very small, we work with standardized residuals as the random variable, i.e.

$$e_t = (x_t - \hat{x}_t \mid x_{t-1}) / \sqrt{\sigma^2}$$

If e_t is ARCH(1), then e_t^2 should behave like a non-Gaussian AR(1) process. Therefore, we can check that ϕ_{11} for e_t^2 should be significant and acf should be exponentially decreasing. The actual values of acf and pacf for e_t^2 showed that AR(1) model is accepted. Hence e_t behaves as an ARCH(1) process. Values of α_0 and α_1 for this data sets were found to be $\hat{\alpha}_0 = 0.82$, $\hat{\alpha}_1 = .18$.

Example 17 *Stochastic Volatility*

In recent years, much attention is placed for GARCH(1,1) models for the stochastic volatility of returns for stock exchanges which we denote as r_t. Stochastic volatility is included by adding stochastic noise term (ε_t) to σ_t as $\sigma_t \varepsilon_t$. This model may be applicable to ore-assay distribution modelling. Thus, we obtain

$$r_t = \sigma_t \varepsilon_t$$

$$\sigma_t^2 = \alpha_0 + \alpha_1 r_{t-1}^2 + \beta_1 \sigma_{t-1}^2$$

where ε_t is the stochastic (Gaussian) white noise.

We can rewrite the first equation as

$$r_t = \exp\left(\tfrac{1}{2} \log \sigma_t^2\right) \varepsilon_t$$

Defining $h_t = \ln \sigma_t^2$; $y_t = \ln r_t^2$, we get

$$y_t = h_t + \ln \varepsilon_t^2 \qquad \text{(observation equation)}$$

where h_t (stochastic variance) is unobserved state process.

The volatility process follows an autoregression as:

$$h_t = \phi_0 + \phi_1 h_{t-1} + w_t, \, w_t \sim \text{GWN}(0, \sigma_w^2) \quad \text{(state equation)}$$

If ε_t^2 is a lognormal distribution, then we obtain a Gaussian state-space model for stochastic volatility and standard DLM could be used to fit the model to data.

Unfortunately, $y_t = \log r_t^2$ is rarely normal, so we keep ARCH normality assumption on ε_t, so long as ε_t^2 is distributed as log chi-squared random variable with 1 d.f., which is given by

$$f(x) = \frac{1}{\sqrt{2\pi}} \exp\{-\tfrac{1}{2}(e^x - x)\}, -\infty < x < \infty$$

and its mean is -1.27 and variance is $\pi^2/2$ (which is highly skewed with long tail to the left).

We simplify the observation equation as

$$Y_t = h_t + \eta_t$$

where η_t is white noise, whose distribution is mixture of two normals, one being centred at zero, i.e.

$$\eta_t = u_t z_{t0} + (1 - u_t)z_{t1}$$

with u_t as an iid Bernoulli process

$Pr(u_t = 0) = \pi_0$ and $Pr(u_t = 1) = \pi_1$, and $\pi_0 + \pi_1 = 1$, $z_{t0} \sim$ iid $N(\mu_1, \sigma_0^2)$, and

$z_{t1} \sim$ iid$N(\mu_1, \sigma_1^2)$.

Then, the likelihood is easily evaluated for $\theta = (\phi_0, \phi_1, \mu_1, \sigma_1^2, \sigma_w^2, \pi_1)'$ with $\pi_j(t)$ independent of t as:

$$\ln L_Y(\theta) = \sum_{t=1}^{n} \ln\left(\sum_{j=0}^{1} \pi_j f_j(t|t-1)\right)$$

where $f_j(t|t-1)$ is approximated by a normal $N(h_t^{t-1} + \mu_j, \sigma_j^2)$ density ($j = 0, 1$ and $\mu_0 = 1$).

Here, h_t^{t-1} is the state prediction given by ARCH model defined earlier.

We can also maximize the likelihood function to estimate the parameters by using the Newton-Raphson method or by EM algorithm to the complete data likelihood (refer Bollerslev, 1986) (see Fig. 5.1).

Example 18   *NYSE Returns*

Here, $y_t = \log r_t^2$ for 2000 daily observations in NYSE shows cyclicity and a large volatility at several points, although it may be locally stationary. A GARCH(1,1) model was fitted with $\pi_1 = \frac{1}{2}$ (fixed) using the quasi-Newton-Raphson method to maximize the likelihood function and the results are as follows: The residuals fitted a $\log(\chi_1^2)$ density very well. This shows that the model for GARCH(1,1) with a mixture of two normals is acceptable for forecasting.

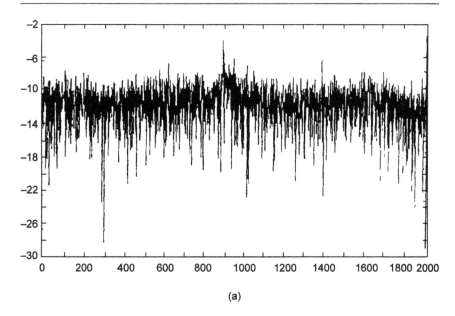

(a)

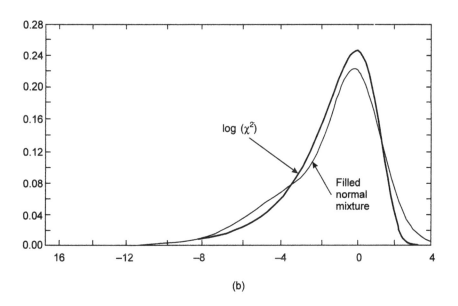

(b)

Fig. 5.1 (a) Volatility as seen in stock market returns for New York stock exchange

(b) Density of log of a chi-square (dark line) and the fitted normal mixture (thin line) for data (a)

Table 5.1 **Mixture of for GARCH (1, 1) normals filled to NYSE data. Parameters ϕ_0, ϕ_1, σ_0, μ_1, σ_1, σw as per example 17 on stochastic volatility**

Parameter	Estimates	Standard errors
ϕ_0	-0.722	0.296
ϕ_1	0.928	0.029
σ_0	1.189	0.066
μ_1	-2.383	0.196
σ_1	2.658	0.099
σ_w	0.224	0.053

5.4.5 Long-memory Models

The acf $\rho(\cdot)$ for an ARMA process at log h converges rapidly to zero, as $h \to \infty$ in the sense that there exists $r > 1$ such that

$$r^h \rho(h) \to 0 \text{ as } h \to \infty$$

Stationary processes with much more slowly decreasing acf, known as a fractionally-integrated ARMA process, or more precisely, as ARIMA(p, d, q) processes with $0 < |d| < 0.5$ satisfy the difference equation of the form

$$(1 - B)^d \, \phi(B) \, X_t \, = \, \theta(B) \, z_t$$

where $\phi(z)$ and $\theta(z)$ are polynomials of degrees p and q, respectively. These processes satisfy $\phi(z) \neq 0$ and $\theta(z) \neq 0$ for all z, so that $|z| \leq 1$, with B as the backshift operator and $\{z_t\}$ is a white noise sequence with mean 0 and variance σ^2. The operator $(1 - B)^d$ is defined by binomial expansion

$$(1 - B)^d \, = \, \sum_{j=0}^{\infty} \, \pi_j B^j$$

where $\pi_0 = 1$ and $\pi_j = \prod_{0 < k \leq j} \dfrac{k - 1 - d}{k}, j = 1, 2, \ldots$

The autocorrelation $\rho(h)$ at lag h of an ARIMA(p, d, q) process with $0 < |d| < 0.5$ has the property $\rho(h)h^{1-2d} \to c$ as $h \to \infty$.

This implies that $\rho(h)$ converges to zero as $h \to \infty$ at a much slower rate than $\rho(h)$ for an ARMA. Consequently, fractionally-integrated ARMA processes prossess 'long-memory'. In contrast, stationary processes whose ACF converges to 0 rapidly, as ARMA processes, are said to have 'short-memory'.

A fractionally-integrated ARIMA(p, d, q) process can be regarded as an ARMA process driven by fractionally-integrated noise, i.e. $\phi(B) \, X_t = \theta(B) \, W_t$ and $(1 - B)^d \, W_t = z_t$.

The process $\{W_t\}$ is called fractionally-integrated white noise and has variance and autocorrelations as:

$$\gamma_W(0) = \sigma^2 \Gamma(1-2d)/\Gamma^2(1-d)$$

$$\rho_W(h) = \frac{\Gamma(h+d)\,\Gamma(1-d)}{\Gamma(h-d+1)\,\Gamma(d)} = \prod_{0<k\le h} \frac{k-1+d}{k-d}, h = 1, 2, \ldots$$

where $\Gamma(\cdot)$ is the gamma function.

The exact autocovariance function of ARIMA(p, d, q) process $\{X_t\}$ can be expressed as:

$$\gamma_X(h) = \sum_{j=0}^{\infty} \sum_{k=0}^{\infty} \psi_j \psi_k \gamma_w(h+j-k)$$

where $\sum_{i=0}^{\infty} \psi_i z^i = \theta(z)/\phi(z)$, $|z| \le 1$ and $\gamma_W(\cdot)$ is the autocovariance function of fractionally-integrated white noise with parameters d and σ^2, i.e.

$$\gamma_W(h) = \gamma_W(0)\,\rho_W(h)$$

The acf $\rho_W(h)$ converges rapidly so long as $\phi(z)$ does not have zeros with absolute value close to 1.

The spectral density of $\{X_t\}$ is

$$f(\lambda) = \frac{\sigma^2 \left| \theta(e^{-i\lambda}) \right|^2}{2\pi \left| \phi(e^{-i\lambda}) \right|^2} \left| 1 - e^{-i\lambda} \right|^{-2d}$$

We estimate the parameters by maximizing Whittle's approximation L_W as follows:

$$-2\ln L_W = n \ln (2\pi) + 2\eta\ln \sigma + \sigma^{-2} \sum_j \frac{I_n(w_j)}{g(w_j)} + \sum_j \ln g(w_j)$$

where I_n is the periodogram, $\sigma^2 g(=f)$ is model spectral density and $\sum_j$ denotes the sum over all non-zero Fourier frequencies, $w_j = 2\pi/n$; $w_j \in (-\pi, \pi]$.

We are not maximizing the exact Gaussian likelihood because it is much more computer intensive.

Long-memory time series are intermediate between ARMA(p, q) (or short-memory) and fully-integrated non-stationary processes in the

Box-Jenkins class. These models are useful in hydrologic series, environmental series such as climate, varve deposits, tectonic uplifts/ subsidence. For more details refer Beran, 1994.

5.5 CASE HISTORY

Earthquakes versus Artificial Explosions

5.5.1 Introduction

In recent years, there has been growing concern about the discrimination of natural earthquakes versus artificial explosions (mining or nuclear) since in the former case, certain hazard precautions and mitigation efforts have to be planned. The seismic recording stations have a continuous record of P and S wave arrivals along two horizontal and vertical directions so that amplitudes and phases are discernible. The amplitude ratios of P and S waves may be less for natural earthquakes as compared to artificial explosions, since the former is for long-time periods (lower frequencies), whereas the latter is for very short-time period (high frequencies). Therefore, spectral analysis of these records may be useful for discrimination, i.e. time series discriminant functions. Although some differences in the spectrum of P and S waves for earthquakes and explosions are present, there is a lot of overlap which indicates that a more comprehensive linear discriminant function may be required. If we use the three directional records, the spectral matrix is

$$\hat{f}(v) = L^{-1} \sum_{l=-(L-1)/2}^{(L-1)/2} X(v_k + l/n) \, X^*(v_k + l/n)$$

which may provide more information along the three geographical directions of the spectrum. Time series discrimination is not restricted to earth sciences (earthquake records) but can be also utilized in other fields such as engineering ('pattern recognition'), climatology, environmental pollution (monitoring and control), etc.

Discrimination of multiple groups based on multivariate data could be of two types:

(a) classification based on mean vectors (μ_i) and covariance matrices (Σ_i) assumed to be unequal and hence, posterior probability

$$p(\pi_i \mid x) = \pi_i \, p_i(x) / \sum_j \pi_j p_j(x); \text{ and}$$

(b) linear discriminant functions based on mean vectors μ_i given all covariance matrices are equal, i.e. $\Sigma_i = \Sigma$ for all groups i,

$$P_i(x) = (2\pi)^{-p/2} \, |\Sigma_i|^{-1/2} \exp\{-\tfrac{1}{2}(x - \mu_i)'\Sigma^{-1}(x - \mu_i)\}$$

so that $\text{LDF} = (\mu_1 - \mu_2)' \, x \, \Sigma^{-1} - \frac{1}{2}(\mu_1 - \mu_2)' \Sigma^{-1}(\mu_1 + \mu_2) + \ln\left(\dfrac{\pi_1}{\pi_2}\right)$

where we classify into π_1 and π_2 if $\text{LDF} \geq 0$ or < 0.

Note: Mahalonobis $D_p^2 = (\mu_1 - \mu_2)' \Sigma^{-1}(\mu_1 - \mu_2)$ and the two misclassification probabilities are equal, i.e. $P(1 \mid 2) = P(1 \mid 2) = \phi\left(-\dfrac{\sqrt{D_p^2}}{2}\right)$.

We use training sets to compute the discriminant functions and the unknown sample can be classified into one of the groups using this established discriminant function.

Numerical Example

Using the logarithms of amplitudes of P and S waves as the inputs (x_1, x_2), we obtain mean vectors for 8 earthquakes and 8 explosions as

$$x_1 = (4.25, 4.95)'; \qquad x_2 = (4.64, 4.73)'$$

with covariance matrices

$$S_1 = \begin{pmatrix} .3096 & .3954 \\ .3954 & .5378 \end{pmatrix}, \ S_2 = \begin{pmatrix} .0954 & .0804 \\ .0804 & .1070 \end{pmatrix}$$

which shows that $S_1 \neq S_2$.

However, to simplify the discrimination, we assume $S_1 = S_2 = S$, where

$$S = \begin{pmatrix} .2025 & .2379 \\ .2379 & .3238 \end{pmatrix}$$

Thus, we obtain

$$\log(p_1(x)) = 22.12 \, x_1 - .98 \, x_2 - 45.23$$

$$\log(p_2(x)) = 42.61 \, x_1 - 16.8 \, x_2 - 59.80$$

So, $\quad\quad \text{LDF} = -20.49 \, x_1 + 15.82 \, x_2 + 14.57$

with $D_p^2 = 14.57$ and probability of misclassification $= 0.04$ (nonsignificant).

This indicates $x_2 - x_1$ is close to the optimal linear discriminant function. So, we can empirically discriminate between earthquakes and explosions on the basis of $\log S$ versus $\log P$ plots without any misclassifications.

The discrimination approach works well when a low-dimensional vector characterizes the groups under study. However, in time series studies with long data sets over a period of time (or spatial coordinates), the characteristic vector dimension could be very large in time domain. A frequency domain approach with only a few discriminating frequencies may still be discernible and are valid even if time-domain covariances matrices are unequal. For such situations, it is convenient to write the Whittle approximation to the log likelihood in the form

$$\ln p_j\left(X\right) = \sum_{0 < v_k < 1/2} [- \ln |f_j(v_k)| - X^*(v_k) f_j^{-1}(v_k) X(v_k)]$$

The quadratic discriminant is of this form

$$\ln p_j(X) = \sum_{0 < v_k < 1/2} [- \ln |f_j(v_k) - |r\{I(v_k) f_j^{-1}(v_k)\}]$$

where $I(v_k) = X(v_k) X^*(v_k)$ denotes the periodogram matrix. For equal prior probabilities, we may assign an observation x into population Π_i, whenever

$$\ln p_i(X) > \ln p_j(X)$$

for $j \neq i, j = 1, 2, ..., g$.

The discriminant series $I = I(\hat{f}, f_1) - I(\hat{f}, f_2)$ for earthquakes and explosions clustering results are summarized below:

Table 5.2   **Values of discriminants (I) for 8 training set data. Setting 0.86 as the discriminant on training set data are perfectly classified.**

	Earthquakes	*Explosions*
1	8.51	0.29
2	0.81	− 2.55
3	30.80	− 1.82
4	2.73	− 1.89
5	7.69	− 1.16
6	21.50	− 2.12
7	20.31	− 2.10
8	15.54	0.93

Table 5.3   **Clusters for above 8 training set data, where earthquake data (4, 5) are misclanified using random clustering method.**

Two Groups	*Cluster I*	*Cluster II*
Random	EQ 123678	EX 12345678
		EQ 45
Hierarchical (partitioned)	EQ 12345678	EX 12345678

The Newton-Raphson algorithm is commonly used for MLE of k parameters $\beta = (\beta_1, \ldots, \beta_k)$ that we wish to minimize with respect to β. We want to minimize $l(\hat{\beta})$, where $\hat{\beta}$ is found by solving $\partial l(\beta)/\partial \beta_j$ for $j = 1, 2, \ldots, k$.

Let $l^{(1)}(\beta)$ denote the $k \times 1$ vector of partials, i.e. $l^{(1)}(\beta) = \left(\dfrac{\partial l(\beta)}{\partial \beta_1}, \ldots, \dfrac{\partial l(\beta)}{\partial \beta_k} \right)$.

Note that $l^{(1)}(\hat{\beta}) = 0$, the $k \times 1$ zero vector.

Let $l^{(2)}(\beta)$ denote the $k \times k$ matrix of second-order partials $l^{(2)}(\beta) = \left\{ -\dfrac{\partial^2 l(\beta)}{\partial \beta_i \partial \beta_j} \right\}_{i,j=1}^{k}$ and assume $l^{(2)}(\beta)$ is non-singular. Let β_0 be an initial estimator of β.

Then, using the Taylor expansion, we have the following expression

$$0 = l^{(1)}(\hat{\beta}) \approx l^{(1)}(\beta_0) - l^{(2)}(\beta_0) \, [\hat{\beta} - \beta_0]$$

Setting the RHS equal to zero and solving for $\hat{\beta}$, we get

$$\hat{\beta}_{(1)} = \beta_0 + [l^{(2)}(\beta_0)]^{-1} l^{(1)}(\beta_0)$$

The Newton-Raphson algorithm proceeds by iterating this result, replacing β_0 by $\hat{\beta}_{(1)}$ to get $\hat{\beta}_{(2)}$ and so on until convergence to $\hat{\beta}$. For MLE $l(\beta) = -\ln L(\beta)$ and any constants can be ignored. In solving this, we replace $l^{(2)}(\beta)$ by $E[l^2(\beta)]$, the information matrix, and inverse of the information matrix is the asymptotic matrix of estimator $\hat{\beta}$, which is approximated by $[l^{(2)}(\hat{\beta})]^{-1}$.

5.5.2 Wavelet Analysis

If time series is stationary, then it makes sense to use the spectral representation theorem that it is composed of stationary cosine/sine waves of various frequencies. Often, time series are, in fact, non-stationary but made stationary by suitable transformation, so that the theory and methods of stationary time series could be applicable for modelling such stationary (transformed) series. However, we may be interested to study the non-stationarity properties of the time series and then the stationarity transformation of the series could be conceptually erroneous. For such non-stationary time series, the interest is on local behaviour rather than global behaviour of the process. For details refer Donoho and Johnstone, 1995.

Example 19

Earthquake and explosion data can be characterized by dynamic Fourier analysis of locally stationary P and S wave data and by shifting the window to cover the entire length of the data. The S component shows power at low frequencies only and the power persists for a long time, whereas for explosion data, we have power at higher frequencies and the power of signals (P and S waves) does not last as long as in the case of earthquakes. We think of the local transformation

$$d_{j'k} = n^{-1/2} \sum_{i-1}^{n} x_t \psi_{j,k}(t)$$

where $\qquad \psi_{j,k}(t) = \begin{cases} (n/m)^{1/2} h_t \exp(-2\pi\, itj/m), t \in [t_k + 1, t_k + m] \\ 0 \qquad\qquad\qquad\qquad\qquad \text{otherwise} \end{cases}$

where h_t is tapered and m is some fraction of n.

Here, h_t is cosine bell taper (means zero) over 256 points and $v_j = j/m$ for $j = 1, 2,\ldots [m/2]$. Thus, transforms are based on local sinusoids.

Another way to analyze the local stationarity of non-stationary processes is through wavelet analysis. We have *father* wavelet ϕ and *mother* wavelet ψ. The father wavelet captures smoothness, low-frequency nature and integrates to unity, whereas mother wavelets capture detailed and high-frequency nature of the data and integrate them to zero. For example, the Harr function is

$$\psi(t) = \begin{cases} 1, & 0 \le t < 1/2 \\ -1, & 1/2 < t < 1 \end{cases}$$

with father wavelet $\phi(t) = 1$ for $t \in [0, 1)$ and zero otherwise.

We can use several wavelet modules such as doublet 4, symmlet 8, etc., which are generated by some numerical methods as they do not possess analytical form.

Since we do not have a periodic function such as cosines/sines, the meaning of frequency, cycles/time is not available, but the scale is important. The orthogonal wavelet decomposition of time series x_t, $t = 1$, 2..., n is

$$x_t = \sum_k s_{J,k}\,\phi_{J,k}(t) + \sum_k d_{J,k}\,\psi_{J,k}(t) + \sum_k d_{J-1,k}\,\psi_{J-1,k}(t) + \sum_k d_{1,k}\,\psi_{1,k}(t)$$

where J is the number of scales and k ranges from 1 to the number of coefficients associated with the specified component. The wavelet functions are generated from the father wavelet $\phi(t)$ and the mother wavelet $\psi(t)$ by transformation (shift) and scaling:

$$\Phi_{J,k}(t) = 2^{-J/2}\phi\left(\frac{t-2^J k}{2^J}\right)$$

$$\Psi_{J,k}(t) = 2^{-j/2}\,\psi\left(\frac{t-2^j k}{2^j}\right), j = 1,\ldots, J$$

The shift parameter is $2^j k$ and scale parameter is 2^j.

The wavelet functions are spread out and shorter for larger values of j (or scale parameter 2^j), and tall and narrow for smaller values of scales j ($1/2^j$ or $1/$scale is the analog of frequency ($v_j = j/n$)).

When $j = 1$, scale parameter is 2 and this is akin to the Nyquist frequency of $1/2$ and when $j = 6$ (scale parameter $= 2^6$) is akin to a low frequency ($1/2^6 \approx 0.016$).

So, larger values of scale refer to slower, smoother (or coarser) movements of the signal and smaller values of scale refer to faster, choppier (or finer) movements of the signal.

The discrete wavelet transform (DWT) of data x_t are the smoothness coefficients $s_{J,k}$ and detailed coefficients $d_{j,k}$ for $j = J, J-1,\ldots, 1$ as follows:

(Smoothness coefficients) $s_{J,k} = n^{-1/2}\sum_{t=1}^{n} x_t \phi_{J,k}(t)$

(Detailed coefficients) $d_{j,k} = n^{-1/2}\sum_{t=1}^{n} x_t \psi_{j,k}(t), j = J, J-1, \ldots,1.$

The magnitudes of the coefficients measure the importance of corresponding wavelet term in describing the behaviour of x_t. The inverse of DWT is denoted as IDWT.

We have x_t data with $n = 2^{11} = 2048$ and DWT is computed with $J = 6$ levels. There are $2^{10} = 1024$ values in $d1$ and $2^9 = 512$ values in $d2$ and so on till $2^5 = 32$ are in $d6$ and $s6$.

Computing the DWTs, the earthquake is best represented by wavelets with larger scale than explosion.

The total energy $= \sum_{t=1}^{n} x_t^2$ and total power in each scale is $(n = 2^{11})$,

$$TP_6^s = \sum_{k=1}^{n/2^6} s_{6,k}^2$$

and $\qquad TP_j^d = \sum_{k=1}^{n/2^j} d_{j,k}^2, j = 1,...,6$

Since the basis is orthogonal, we obtain

$$TP = T + TP_6^s + \sum_{j=1}^{6} TP_j^d$$

which is listed in the Table 5.4.

Table 5.4   **Smoothness and detailed coefficients of DWT for eathquakes and explosion**

Component	Earthquakes	Explosions
s6	0.009	0.002
d6	0.043	0.002
d5	0.377	0.007
d4	0.367	0.015
d3	0.160	0.559
d2	0.040	0.349
d1	0.003	0.066

Nearly 80% of total power of earthquakes is explained by higher scale details d4 and d5, whereas 90% of total power of explosions is explained by smaller scale details d2 and d3. Thus, S-waves for earthquakes show power at high scales (or low 1/scale) only, and power remains strong for a long time. In contrast, the explosions show power at smaller scales (or higher 1/scale) than the earthquake, and the power of the signals (P and S-waves) do not last as long as those for the earthquakes. Therefore, wavelet transforms can be used for discriminating explosions and earthquakes.

5.6 EXERCISES

1. Find the state-space representation for MA(1) process
   $X_t = z_t + \theta z_{t-1}$ and for MA(2) process $X_t = z_t + \theta_1 z_{t-1} + \theta_2 z_{t-2}$
2. For AR(2) process $X_t - \phi_1 X_{t-1} - \phi_2 X_{t-2} = z_t$, the state vector is represented as $\alpha'_t = (X_t, X_{t-1})$

   Show that $G = \begin{bmatrix} \phi_1 & \phi_2 \\ 1 & 0 \end{bmatrix}$.

3. Find a state-space representation of AR(2) process with state vector $\alpha'_t = \left[X_t, \hat{X}(t-1) \right]$, where $\hat{X}(t-1)$ is the optimal one-step-ahead predictor of X_{t+1}

   Ans: $G = \begin{bmatrix} 0 & 1 \\ \phi_2 & \phi_1 \end{bmatrix}$ with $w'_t = (1, \phi_1)z_t$.

4. Suppose $\{X_t\}$ is AR(2) with

   $$X_t = \phi_1 X_{t-1} + \phi_2 X_{t-2} - z_t, \ \{z_t\} \sim WN(0, \sigma^2)$$

   We observe 7 data symmetrically about Y_3, which is missing. Find the best estimator of Y_3 given the observed Y values and the model parameters.

   Ans: $\hat{Y}_3 = [\phi_2 (Y_1 + Y_5) + (\phi_1 - \phi_2 \phi_1)(Y_2 + Y_4)]/(1 + \phi_1^2 + \phi_2^2)$

5. Let $\{X_t\}$ be a linear process with $\{z_t\} \sim \text{IID}(0, \sigma^2)$ and $\eta = Ez^3$ Show that the 3rd order cumulant function of $\{X_t\}$ is $C_3(r, s) =$

   $$\eta \sum_{s=-\infty}^{\infty} \psi_i \psi_{i+r} \psi_{i+s}$$

   If $\{X_t\}$ is Gaussian linear process, then show that $C_3(r, s) \cong 0$ and $f_3(w_1, w_2) \cong 0$.

6. For a convergent ARCH(1) process, we have

   $$X_t = z_t \sqrt{a_0 \left(1 + \sum_{j=1}^{\infty} \alpha_1^j z_{t-1}^2 \ldots z_{t-j}^2 \right)}$$

   where $E X_t^2 = \alpha_0/(1 - \alpha_1)$ and $0 < \alpha_1 < 1$, $\alpha_0 > 0$.
   Show that $EX_t^4 < \infty$ iff $3 \alpha_1^2 < 1$.

7. Consider a system process given by $x_t = -.9x_{t-2} + w_t$, $t = 1, 2, \ldots, n$ where $x_0 \sim N(0, \sigma_0^2)$ and Gaussian $W_t \sim (0, \sigma_w^2)$.
   This system is observed with noise $y_t = x_t + v_t$, where v_t is Gaussian white noise $v_t \sim N(0, \sigma_v^2)$
   Also, $x_0, \{W_t\}$ and $\{v_t\}$ are independent processes.
   (i) Write the system and observation equations in the form of a state-space model
   (ii) Find σ_0^2, which makes the observation y_t stationary.

8. Let the state conditions be $x_0 = w_0$ and $x_t = x_{t-1} + w_t$ with observations $y_t = x_t + v_t$, $t = 1, 2, \ldots$ where w_t and v_t are independent, Gaussian white noise processes with variances σ_w^2, σ_v^2, respectively.
   Show that y_t is ARIMA(0,1,1) or IMA(1,1).

9. Assume a third degree polynomial in time to yearly average global temperature differences in °C, i.e.

$$y_t = \beta_0 + \beta_1 t + \beta_2 t^2 + \beta_3 t^3 + \varepsilon_t$$

where $\{\varepsilon_t\} \sim WN(0, \sigma^2)$.

Write this model in state-space form given $y_t = x_t + v$ with $\nabla^3 x_t = w_t$ and w_t and v_t are independent white noise processes.

10. Suppose AR(1) process $x_t = \phi x_{t-1} + w_t$ with $W_t \sim (0, \sigma_w^2)$ has a missing value at $t = m$, so $y_t = A_t x_t$, where $A_t = 1$ for all t (except at $t = m$, where $A_t = 0$).

Let $x_0 = 0$ with variance $\sigma_w^2/(1 - \phi^2)$.

Show that the Kalman smoother estimators are

$$x_t^n = \begin{cases} \phi y_1, & t = 0 \\ \left(\phi/(1 + \phi^2)\right) (y_{m-1} + y_{m+1}) & t = m \\ y_t & t \neq 0, m \end{cases}$$

with mean square covariances

$$\rho_t^n = \begin{cases} \sigma_w^2 & t = 0 \\ \sigma_w^2/(1 + \phi^2) & t = m \\ 0 & t \neq 0, m \end{cases}$$

References

1. Akoi, M., 1987, *State-Space Modelling of Time Series*. Springer, Berlin.

2. Beran, J., 1994, *Statistics for Long-Memory Processes*. Chapman and Hall, New York.

3. Bollerslev, T., 1986, Generalized Autoregressive Conditional Heteroscedasticity, *J. Econometry*, 31: 307-327.

4. Donoho, D.L., and Johnstone, I.M., 1995, Adapting to unknown Smoothness via Wavelet Shrinkage, *Jour. Amer. Statistical Assocn.* 90: 1200-1224.

5. Gouriéroux, C., 1997, *ARCH Models and Financial Applications*, Springer-Verlag, New York.

6. Jones, R.H., 1984, Fitting Multivariate Models to Unequally Spaced Data. E. Parzen (Ed.), Time Series Analysis of Irregularly Observed Data. Lecture Notes in Statistics, 25, Springer-Verlag, New York.

7. Tsay, R., 1987, Conditional Heteroscedasticity in Time Series Analysis, *Jour. Amer. Statistical Assocn.*, 82: 590-604.

8. Tong, H., 1983, Threshold Models in Time Series Analysis, Lecture Notes in Statistics, 21, Springer-Verlag, New York.

SUBJECT INDEX